FUNDAMENTALS OF BEEKEEPING

FUNDAMENTALS OF BEEKEEPING

Dr. T.V. Sathe

Entomology Division,
Department of Zoology,
Shivaji University, Kolhapur-416 004

2024

Daya Publishing House®

A Division of

Astral International Pvt. Ltd.

New Delhi – 110 002

First Published: 2006
Reprinted: 2024
ISBN 978-81-7035-420-8

Published by : **Daya Publishing House®**
A Division of
Astral International Pvt. Ltd.
– ISO 9001:2015 Certified Company –
4736/23, Ansari Road,
New Delhi - 110 002
Phone: 011-4354 9197, 2327 8134
E-mail: info@astralint.com
Website: www.astralint.com

Laser Typesetting : **Vaishnav Graphics & Systems**
Delhi - 110 055

Digitiality Printed at : **Replika Press Pvt. Ltd.**

Dedicated to

my mother

Nakula

PREFACE

Honey bees are important tool of commercial industry, the apiculture. Therefore, much more literature is available on applied aspects of bee keeping (Smith, 1963; Rodionov & Shabarshow, 1986; Ghosh, 1994 etc.). However, very little is known about the details of basic aspects of honeybee such as anatomy, morphology, biology, etc. except the work of snodgrass (1956). Hence, credit goes to snodgrass for anatomical studies of bees. In the present text international status of beekeeping, agricultural importance, history of apiculture, external morphology of adult bee and its immature stages, anatomy of adult bee and its immature stages, physiology, biology and behaviour have been described with suitable sketches and figures.

Since apiculture has commercial base, University Grant Commission, New Delhi, insisted apiculture in syllabi at graduate and post graduate level and at diploma courses. The basic aspects of bee covered in this book are extremely important for filling the gap of such scanty literature and from view of understanding the life mechanism and style of the bee and further adopting this knowledge for increasing reproductive potential and production of honey, wax and services of bees for agriculture purpose as pollinator for increase in yield of crops.

I gratefully acknowledge the reference work of various writers and I must thank my daughter Asawari (MBBS III) for her bee like efforts in sketching the figures of the book and other help rendered for this manuscript. I hope that this book will fulfil the requirement of basic literature on honeybee and will be helpful for students, teachers and apiculturist in India and abroad.

T.V. Sathe

Contents

1.
Introduction

Biotechnology is gaining more importance in various biological sciences such as Agriculture, Horticulture, Forestry, Food technology, Medicine, Cosmetics and industries like silk, fabrication, lac, biopesticides, biofertilizers, biological pest control agent production technology, etc because of various kinds of industrial and pesticidal pollutions. The demerits of pesticidal pollution are very crucial from the view point of survival of human being and other biota. Therefore, any production technology must have its own merits and should be eco-friendly. We need honey for better health and wax for preparing various goods used in scientific work and day to day life. Natural resources or products are always better than the artificial ones. Therefore, they should be managed for the betterment of human being and nature. Honey is natural product prepared by honeybees from pollen and nectors and their own secretions and very useful in health care of man.

The collection of honey from honeybees by beekeeping is refered as Apiculture. The business of apiculture not only gives us honey but also an economical product *i.e.,* wax which has more than 200 uses in day to day life and scientific laboratories, etc.

Indian farmers are largely involved in increasing the yield of the crops in their farms by using the knowledge of plant protection, water management, reliable modern methods of agriculture and by using biotechnology. In spite of investing large amount, they could not get the expected yield of crops and quality seeds. It has several reasons but, scarcity of insect pollinators is main reason. Honeybees are very good pollinators of agricultural and non-agricultural plants. Thus, play a very important role in modern agriculture for increasing the yield of crops, several times more than control.

Cutting of trees and bushes remove the shelters of honeybee because of which the population of bees and pollination of crops is reduced. It is experimentally proved that 85% of pollination is brought by honeybees alone in various agroecosystems. It is also clearly indicated that agricultural benefits are more than the combine benefits of the products, honey and wax they produce. Thus, apiculture business is very wonderful job for agriculturist and industrialist for obtaining the benefits.

Apis indica and *Apis mellifera* bees are easily rearable by adopting wooden boxes. This fact is not known to Indian farmers. Therefore, they are not taking any interest in beekeeping. Developed countries are practing beekeeping business for increasing the yield of agricultural and horticultural crops since last many decades. Indian farmers should consider beekeeping with advanced methods which will be helpful for increasing the yield of crops in addition to regular benefits obtained from honey and wax. Indian agriculture will not increase the yield of crops on large scale without beekeeping in the fields.

In India, 2.5 lakh of hectre of land is under hills and forestry which is useful for beekeeping. Now-a-days, in India, 45000 villages are involved in Apiculture. It is estimated that more than 14 lakh of bee colonies are domesticated in various parts of India and those can be increased upto 3 crore. Because, 36 lakh hectre of land is under cultivation of oilseeds crops, horticultural crops, pulses and cereal crops, etc. in India which will provide good source of pollens and nectors for honeybees. Thus, crops would be pollinated very potentially by bees as they visit crops frequently.

Honeybee population is threaten due to excess cuttings of trees and shrubs, use of pesticides in agro and forest ecosystems and destruction of wild honey combs, etc. A scientific knowledge of beekeeping is therefore, essential for successful beekeeping business. One should know the recent methods of beekeeping and primitive and crude methods of apiculture should not be entertained as they affect the population of bees by killing eggs, larvae, pupae and adult bees during harvest of honey from combs.

At present several countries of the world are greatly involved in apiculture business. The most leading countries in apiculture business refers to USA, Canada, Australia, New Zealand and

Mexico. African and European countries are also largely involved in beekeeping as it is very profitable. Anyone can start beekeeping. Many people of the world keep few hives for obtaining honey for their families and beloved persons. While, others can manage 200 to 1000 hives from the view point of business. In Mexico, a single person can manage more than 50,000 hives for beekeeping business and this is highest record in the world. In certain parts of the world beekeeping is practiced for pollination of plants in addition to collection of honey and wax.

Historical Account

International

During the period of 218 to 201 BC apiculture was most popular in Italy specially Rome province. In Greece, the King Pericles hopefully implemented beekeeping practices around 429 BC. In ancient Egypt beekeeping was practiced during 663 to 610 BC which is evidenced by carvings on pyramid. In Mayans and Aztecks also apiculture was well known practice for culturing honey. In most of the above cases, they used earthen pots, cane, bamboo, baskets, mud hives, straw hives and ropes, etc. They sequeezed honey out from the wax hives mostly after the usual flowering of the seasons for protecting honeybees. Thus, honey was extracted only once during the year. As regards to the methods of beekeeping almost every country attempted primitive methods. However, Langstroth (1851) for the first time discovered the movable hive and scientific method of beekeeping.

In European countries such as Germany, England and France modern methods, have been adopted during the period of 1655 to 1809 A.D. They tried various types of boxes for apiculture. They review the beekeeping practices by adopting several experiments and alterations in equipments. Hence, this period is considered as evolutionary period of beekeeping. In Russia, honeybees were successfully domesticated for apiculture in pokropovic province. Russians for the first time started the school of beekeeping in Gurbania. Beekeeping in American subcontinent was also in full form during those days. It is evidently known that 50 different kinds of hive boxes have been tried for beekeeping in Europian and American countries.

American scientist Langstroth is father of apiculture who designed hive box with frames. The hive box was specially useful for extraction of honey by protecting the lives of honeybees, in all stages *i.e.*, egg, larva, pupa and adult. Langstroth investigated the modern hive in 1851 AD. However, the first honey extractor was investigated by Major Von Hruska in 1865 AD. It is very interesting that the recent honey extractors are based on the principle adopted by Von Hruska. Likely, Johanns Mehring invented the artificial comb foundation sheet for the first time in 1857 AD. In Belgium excited and infuriated swarms of honeybees were used as a weapon in the world war-I for stringing army force.

Indian History

Man had started the use of animal products since time immemorial and honey has been used since prehistoric period as reported in Vedas, Puranas, Ramayana, Mahabharat and Charak Sanghita, the religious literatures of India. As regards to honey-bees, there are several figures of honeybees on crowns and maces of kings. Figures and caryings of honeybees, combs and hives are also found on coffins, tombs and coins of ancient empires. In India, honey was well known since epic era. In Ramayana it has been described that the mandhuban was maintained for taming bees for culture of honey for sugriva, the king of Kiskindya.

In India, a foreign person Doglas, from England started beekeeping as a modern technique in 1884 AD. This fellow was serving in Post and Telegraph department. Because of his tremendous interest in beekeeping, he, for the first time contributed for the book, "Hand Boook of Beekeeping" from India. Beekeeping boxes were brought for usual practice in Shimla in 1884 from England. Similarly, imported beekeeping boxes have been tried in 1907 AD at Poona.

In Tamil Nadu, Fr. Newton prepared beekeeping hive locally and utilized in beekeeping business as a modern device. He was serving in St. Joseph college, Trichy. Agriculture college coimbatore is first institute which started teaching cource in beekeeping. In Kanyakumari district of Tamil Nadu apiculture business was more famous and in full form due to the encouragement of Christian missionaries and YMCA dignities. The above authorities also boosted beekeeping in Kozhikode province of Kerala. From Karnataka beekeeping was practiced with increased attension in Coorg.

Beekeeping in Orissa is performed individually and also on cooperative basis. Gopabandhu Chowdhury and his son

Manmohan made tremendous efforts for popularizing apiculture in Cuttack district. They practiced beekeeping in their Gandhian Ashram situated at Bari (Cuttack). In Oriya language Manmohan Choudhury wrote a book on apiculture for the first time. Through cooperative movement, Sahasrabuddhe expanded this business in Koraput district. Ganjam and Balasore districts of Orissa are quite famous for cooperative activities of apiculture.

Beekeeping was practiced on large scale on the cooperative basis in various districts of West Bengal as a part of post independence programme. 24 Paraganas district beekeeper's cooperative society Ltd., run several cooperative programmes on apiculture and thus, became model in India. This society gave fillip to several cooperatives and beekeepers in the state. Other agencies like Sundarban Khadi and VI Association and Ramakrishna mission successfully practiced beekeeping at Canning and Navendarpur respectively in West Bengal. In addition to this Mrinmoy Das established organizations like Human and Modern Agriculture for doing research in apiculture and agriculture and marketing of flora oriented specific honey in West Bengal.

Rawat Apiaries took organized efforts in beekeeping in North India. As a result, apiculture practices expanded from Shimla to Ranikhet. Now a days apicultural activities are in peak in this region. Production of uniflora honey, supply of standard appliances and literature for beekeeping are important activities of Rawat Apiaries in Uttar Pradesh and rest part of the country. Credit goes to Rawat Apiaries for producing uniflora honey for the first time in India.

In Bihar, special attempts have been made by Zilla Khadi Gramodyog Sangh, Muzzarfarpur for popularizing beekeeping. As a result, other organization started beekeeping on large scale in Bihar.

In Assam, beekeeping is practiced individually and on cooperative basis. Ramkrishna mission started apicultural practices in Meghalaya and Arunachal Pradesh and gave fillip to other organizations and agencies to initiate beekeeping in above states.

In Mizoram and Nagaland, apicultural activities are sponsored under rural development programme while in Manipur Khadi and VI Association played important role in starting beekeeping activities. Now, apicultural activities are taken up as a Govt. programme in the state.

As regards to research in apiculture, Punjab Agricultural University is actively enganged in increasing the yield of honey to large extent by introducing improved variety of *Apis mellifera* in Punjab region. Likely, central Bee Research Institute Pune, also doing research in beekeeping. Similarly, ICAR is also involved in "Lab to Land" programme in apicultural research. Shivaji University Kolhapur is going to start apiculture teaching course under diploma in Insect Biotechnology which will be more helpful to Western Maharashtra as Western Ghats is very important part for pollens and nectors for bees. However, we must thank to Khadi and village industries commission and state khadi and village Industries boards for their sincere efforts made for initiating and expanding apicultural activities in India.

Apicultural practices have very bright future because there is tremendous demand to bee products viz., honey and wax and bee services in modern agriculture for increasing the yield of crops at global scenario. Hence, National and International Organizations like IBRA, DST, ICAR, INSA, KVIC, FAO, WHO etc. are coming forward for boosting apiculture by providing services and solving the problems related to apicultural activities by organizing seminars, workshops, conferences etc.

2.
BIODIVERSITY OF HONEYBEES

Honey producing bees belongs to the class-insecta of order Hymenoptera. The taxonomic position of honeybee is given below:

Phyllum	–	Arthropoda
Class	–	Insecta
Order	–	Hymenoptera
Family	–	Apidae
Genus	–	*Apis*

Honeybee Species Diversity in India

1. Rock bee - *Apis dorsata* F.
2. Indian bee - *Apis cerena indica* F.
3. Little bee - *Apis florea* F.
4. European bee - *Apis mellifera* F.
5. Italian bee - *Apis mellifera ligustica*
6. American bee - *A. m. carnica*
7. Russian bee - *A. m. caucasia*
8. Britan bee - *A. m. lehzeni*
9. African bee - *A. m. adansoni*
10. Chinese bee - *A. m. sinensis*
11. Japanese bee - *A. m. japonica*
12. French bee - *A. m. mellifera*
13. Indian bee - *A. m. indica*

There are about 20,000 species of honeybees in the world. All honeybees belong to order Hymenoptera. Honeybees are scattered in three families namely, Apidae, Bombidae and Meliponidae of order Hymenoptera. The family Apidae is the real family which produce honey in large quantity through honeybees and the products are manageable in beekeeping business. Other families have less importance in beekeeping. The family Bombidae is mainly scattered in temperate region and the bees have separate waxen pots for keeping their broods and honey separately. The individual of the family produce less honey. The family Meliponidae is characterized by having no sting and produce less honey. Therefore, commercially not exploitable. However, some species like *Melipona beechii* are reared in South America and Mexico and *Trigona staudingeri* in Africa. The honey of these bees have pleasant taste and thus, special one.

Apis dorsata F. (Fig. 1)

A. dorsata is called by different names *viz.*, Giant honeybee, Rock honeybee, Sarang bee and Bombara bee. *A. dorsata* is largest and dark brown in colour which measures about 20 mm in body length. It has larger sting. These species are distributed all over India from hilly regions of about 1200 m to the plains. They prepare largest hive measuring about 6 ft long and 3 ft wide and found attached to rocks and tall trees. The comb is verticle, wider at top and tapers to bottom. The top regions is faciliated for storing honey and pollens and bottom shows various cells including queen cell,

Fig. 1 : Adult : *Apis dorsata* (worker)

egg cells and larva cells. Single comb can produce an amount of honey from 36 to 100 kg. Several hives of this bee are located on single tree ranging from 70 to 156 hives. The workers are very active creatures whcih pollinate more than 12,000 flowers per day. *A. dorsata* migrate for avoiding extreme adverse climate and for searching pollens and nectors. They have feroceous behaviour hence, practically difficult and impossible to domesticate in beekeeping business. However, it is observed that *A. dorsata* can stay for a preiod of 5 to 6 months at single place. On an average more than 40,000 bees are found in each comb. They have ability to kill person through sting. As this species is very very potential in producing honey on largest quantity than other bees, there is need to have break through from the view point of its domestication by research work. *A dorsata* species are very potential pollinators of agricultural and horticultural crops. Farmer destroy the combs of this bee due to fear and the feroceous stinging behaviour of this bee, that affect pollination and decrease the yield of crops. These bees should be managed for agriculture purpose for increasing the yield of the crops. This fact should be explained to Indian farmers.

Apis cerena indica

It is commonly known as Indian bee and distributed in South East Asia, China, Japan, Indonesia Phillipines, Pakistan, Sri Lanka, Nepal, Bangladesh etc. Indian bee is the only easily domesticated bee in India. It is medium sized, measuring about 12 to 14 mm in body length and 6 mm in width. It is yellowish brown in body colouration. It prepare parellel combs on trees, hollows of trees, and varities of cavities in the environment, in dark places such as broken pots, old buildings etc. *A.c. indica* construct number of combs at single place and live their for many generations. The comb has mainly two regions. The top region contain strong honey and pollens while, the bottom region contain various cells, called brood comb. On the brood comb queen cell, drone cells, and brood cells are located. The queen lay eggs in brood cells for development of new bees. Indian bee has two varieties namely, Gandhiana and Indica. Gandhiana is restricted to hilly region and is darker and larger than Indica, and collect more honey than Indica. While, Indica is found in plain region of our country. It is smaller than Gandhiana and yellowish in body colouration. As regards to the research on this bee, cross breeds/ hybrids have been tried, as a result, a hybrid honeybee of larger size is released in West Bengal which produces larger quantity of honey. This hybrid is cross breed of *Apis cerena* hill type and plain type. *Apis cerena* is European bee, found in Europe and larger than Indian species. *Apis indica* (Indian bee) can produce 3.6 to 4.5 kg and 1.3 to 2.2 kg honey in hilly and plain regions respectively in India. However, swarming and absconding are draw backs of this bee. Therefore, wax moth (*Galleria mellonella*) is limiting factor as a natural enemy of bees. The most important thing regarding this bee is that it is very gentle and easily domesticated in India. In a single coloney they are 15000 to 20,000 in number. Annually, these bees can produce 4 to 10 kg of honey if sufficient flora is available. As like *A. dorsata*, this bee is also found acting as potential pollinator of agricultural crops. Its hive is again characterized by having minimum six combs constructed parellel to each other. Sometimes worker bees of this species can sting if excited by man. However, their sting is not so serious. Indian bees should be protected from crude methods of beekeeping by adopting scientific modern methods. These bees should be protected and encouraged in order to produce more benefits by their products and pollinating services.

Apis mellifera

It is called as European bee. It has origin from Italy and now introduced in several countries of the world. Its characteristics and behaviour is very similar to *A.c. indica*. There are several subspecies of *mellifera* scattered in several countries as mentioned in diversity of honey bees. They prepare the combs in enclosed spaces and are parallel. In each coloney, number of bees is about 500 and average honey yield is 45-181 kg per colony per year in America. This species is quite adaptable for domestication every where. Therefore, it is favourite bee of the world. In Africa this species yielded maximum 959 kg honey while, in some other countries about 450 kg honey is yielded. The queen is prolific, swarms are less and bees are dosile in nature. Out of several varieties of *A. mellifera*, Italian variety is domesticable everywhere in Europe and America. Thus, it is visualized as best species for beekeeping in the world. From a single colony about 10-15 kg honey is produced per year. In a single colony more than 30,000 bees are observed in this case. The bee is more fecund, and thus, very potential pollinator of agricultural crops. In Latur and Osmanabad districts of India, sunflower is widely cultivatd. The bees can collect large amount of pollens, and bring about the pollination of this crop and thus increase the yield by six-times than the control. European bees should be exploited on large scale in India for collecting honey and wax and for pollination services.

Apis florea

It is called little bee. It is characterized by having body colour brown with white stripes. It measures about 6 mm in body length and 3 mm in width. Thus, it is smallest of all the bees. It prepare single vertical comb measuring about 15-45 cm long and 15-30 cm wide. It provides only 200-500 gm of honey per year. Due to its nomadic feature it is not rearable for apiculture. *A. florea* species are abundant in plain regions of our country but, very interestingly not found above 450 m. The sting of this bee is painful. Its combs are palm sized and located on bushes, hedges, caves, buildings etc., as suspended bodies. These bees should not be destroyed and killed at the time of harvest of honey from natural hives, because, they also play an important role in pollination of agricultural crops and thus, responsible for increasing the yield of crops.

Dammer bee

Dammer bee is black in colour and measures about 3 mm in body length and 1 mm in width. It is having no sting. Its comb is very small, about the size of grape yielding just one or two tea spoons of honey per year. Its honey is aromatic and sour in test. This species is not useful in beekeeping business because of non economical value of honey.

3.

External Morphology of Honeybee

The body of honeybee is divisible into head, thorax and abdomen (Fig. 2). The head is composed of six segments. The segments are fused together forming a compact head capsule. The head consists of a pair of compound eyes, three ocelli or simple eyes on top of the head and a median ocellus in front of other two, a pair of antennae and typical chewing and lapping type of mouth parts. The cranium like head is separated from thorax by a distinct neck. The head contains brain, the suboesophageal ganglion and anterior parts of the alimentary canal. The shape and external features of the head of worker bee are shown in Fig. 5.

The head is triangular, narrow and pointed from below where mouth parts are appended. It is more flattened antero posteriorly from side view. On the lateral angles of the head there are two large compound eyes. The face is convex and rear surface is more or less concave to which anterior end of thorax is snugly fitted. Antennae arise from the face between the lower halves of compound eyes and comes close together with anterior part. The basal segment of antennae is fixed with small circular membranous sockets of headwall. The labrum is attached to face from its lower edge. The labrum is broad and flap like to which a pair of muscles is attached from the facial wall of the head. The functional mouth of bee is located behind the labrum. Vertex is top part of the head. The frons are located above the face and between the compound eyes. Below the eyes are genae. Clypeus is a large shield like plate which is located at facial area below antenna. The limiting groove is epistomal sulcus. Clypeus is extended upward between the genae into the area of the frons. Important muscles of the clypeus are attached to sucking organs in the head region. A pair of lateral pits are located to anterior posterior of clypeaus and marks the anterior roots of the endoskeletal tentorium of the head.

The hind wall of the head capsule has a perforated pentagonal hole called the occipital foramen, through which oesophagus, dorsal blood vessel, nerve cord, air tubes and salivary ducts are passed out. Occiput lies in the area of the head wall above and sides of the foramen. Post genae is the lower lateral area of foramen. Surrounding the foramen is post-occipital sulcus. Below this is narrow post occiput which encloses foramen. Two slit like depressions are the posterior roots of the tentorium which are located in the lateral part of the post occipital sulcus. Proboscis fossa is 'U' shaped depression of sclerotic wall of cranium which is below the foramen. From the membranous floor of proboscis fossa proboscis is suspended.

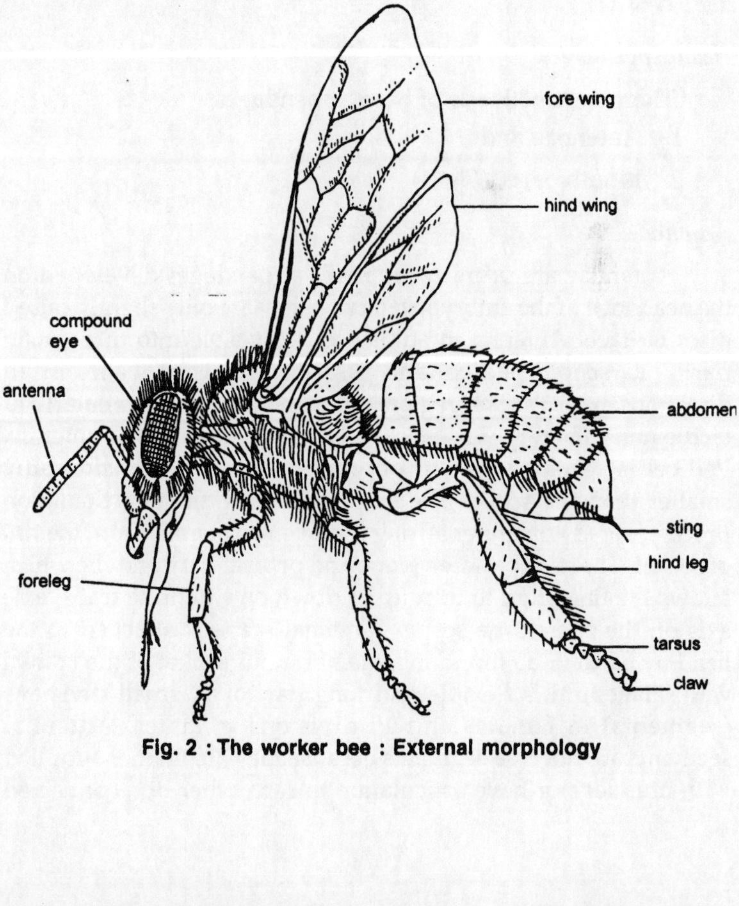

Fig. 2 : The worker bee : External morphology

Characteristics of head in different castes

Queen	Worker	Drone
More rounded on the sides and little wider in proportion than the length.	Head elongated than queen	Much longer than that of either of the female forms and nearly circular.
(Fig. 3a)	One pair of brood food gland present (Fig. 3b)	Greatly enlarged compound eyes which are contiguous on the vertex.
		Ocelli displaced downward on the face by union of compound eyes above them. (Fig. 3c)

The Head

Head Appendages

There are two kinds of head appendages

1. Antennae and
2. Mouth parts. (Fig. 4)

Antennae

Antennae are primitive cephalic appendages developed on the head lobe of the embryo. In larvae, they are only slightly raised discs on face. Antenna in adult bee is divisible into three main parts *i.e.,* scape, pedicel and flagellum. It is movable in all directions except socket region and is sense organ particularly of tactile and olfactory sense. Scape is rigid basal stalk of antenna. Pedicel is second segment of antenna which is considerably smaller than the scape. The scape is having single articulation process, called antennifer arising from ventrolaterally from the rim of socket. The joint between scape and pedicel is dicondylic which allows the flagellum to turn up or down on a definite transverse axis on the end of the scape. Antennal scape is attached to the head by basal knob found in a membranous socket of the cranial wall. Flagellum is flexible and long part of 11 small divisions (segments) in females and 12 divisions in males. Antennal segments are not true segments because they are neither supplied with muscles nor have articulation to each other but, connected

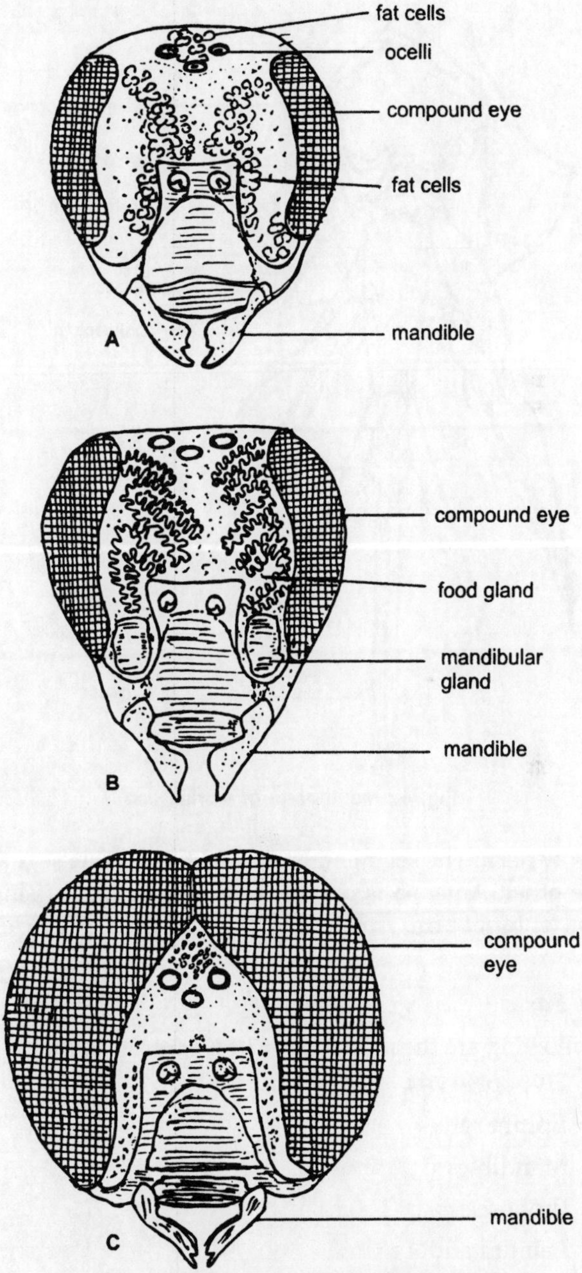

Fig. 3 : (a) Head : queen; (b) Worker; (c) Drone

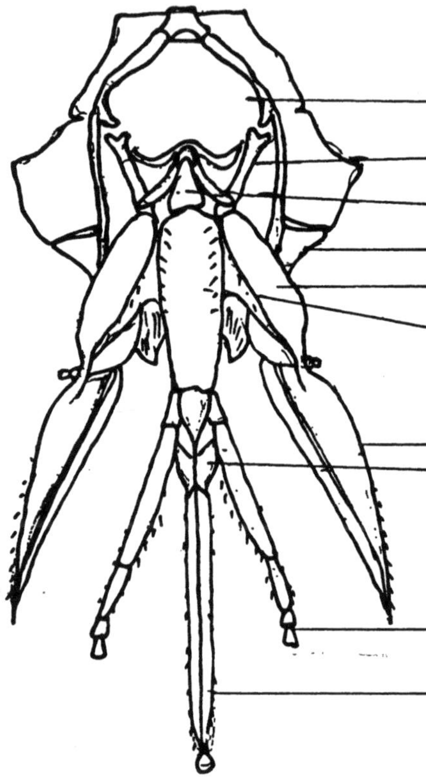

Fig. 4 : Mouthparts of worker bee

by narrow neck. The segmental rounded base of one segment fits into the other. Antenna is provided with 4 muscles arising from anterior tentorial arm. Antenna of honeybee has a typical shape (Fig. 6).

Mouth Parts

Following are the mouth parts of worker honeybee.

1. Hypopharynx
2. Epipharynx
3. Mandibles
4. Proboscis
5. Labium and
6. Maxillae

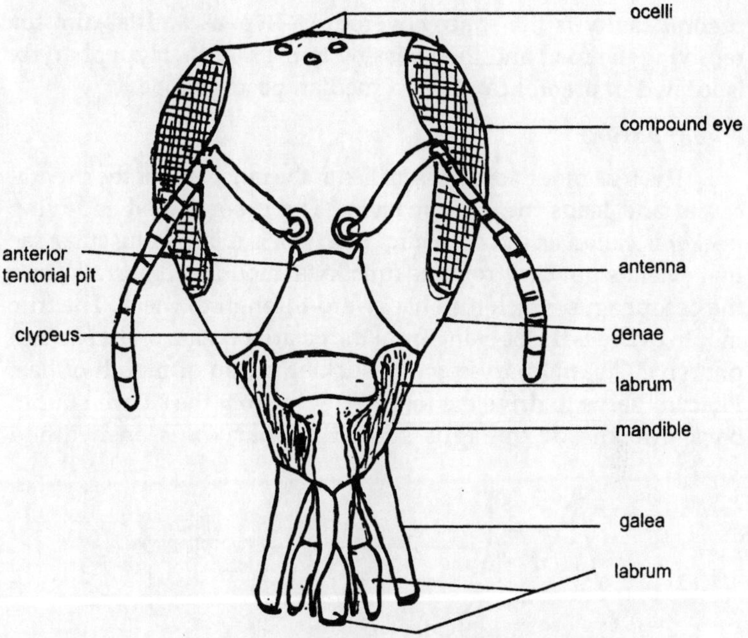

Fig. 5 : Head of worker bee

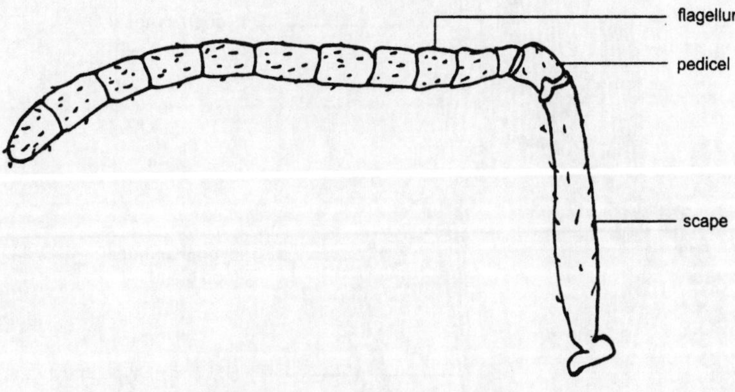

Fig. 6 : Antenna of honeybee

Preoral cavity, hypopharynx, mandibles, maxillae and labium are the important feeding apparatus of honeybee. Mandibles, maxillae and labium are postoral segmental appendages. The labium is formed by the fusion of a pair of second maxillae. The preoral cavity is the space covered by clypeus and labrum for receiving the food and then passing to the mouth. Hypopharynx is located in preoral cavity as a median postoral lobe.

Hypopharynx

The true mouth of an insect lies at the inner end of the preoral cavity and leads into the pharynx. The preoral food receiving pocket is called as the cibarium. Pharynx is a large muscular sac that extends upward from the functional mouth and narrows into the oesophagus which turn backward through the neck. The true mouth of bee is the opening from the cibarial region of the sac into pharynx. Cibarial part is active sucking organ of mouth of bee. Pharynx serve to drive the food received from the cibarial pump back into the oesophagus since the pharynx is ensheathed

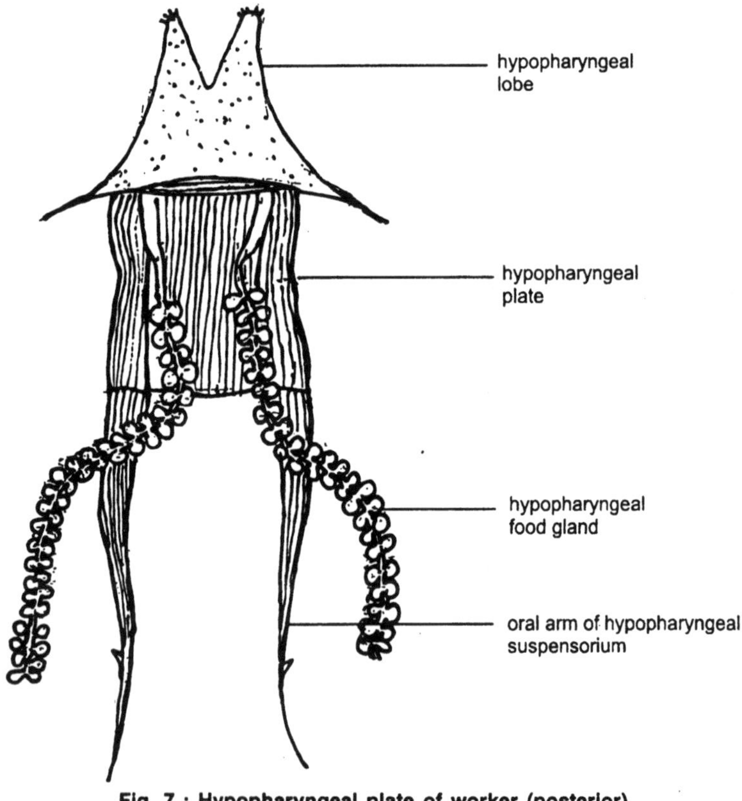

Fig. 7 : Hypopharyngeal plate of worker (posterior)

in layers of circular and longitudinal muscle fibers and are strongly contractile. Hypopharynx of bee larva is united with the labium in the median lobe of mouth parts. Silk duct opens at the tip of the labium.

The Hypopharyngeal glands (Fig. 7)

The food glands of honeybee are refered as hypopharyngeal glands but not pharyngeal. Schiemenz (1883), Soudek (1927), Kratky (1931), etc. described the structure of hypopharyngeal glands of honeybees. "Royal Jelly" or bee milk is secreted by worker bees which is fed to the larvae, queen and drones in the comb. This secretion is secreted by long gland found inside the head. This gland is paired one and with many loops and coils which opens by ducts on the suboral plate of the hypopharynx. In the worker, the food gland consists of numerous small, oval cellular bodies with short neck to an axil duct (Fig. 7). In drones the hypopharyngeal gland is also more or less similar in structure *i.e.*, like a branch of grapes. Hypopharyngeal glands are generally found in Hymenoptera and not in other orders of insects. These glands are large in worker but may be present in drones and queens.

The Epipharynx (Fig. 8)

The epipharyx is a pad like and soft structure. It is triangular in shape from behind and with a thin median lobe, like a keel from the posterior side. From the inner wall of clypeal region of the head is a lobe projecting behind the labrum is called as the epipharynx.

There is an oval elevation having small sense hairs on each sides of the keel. The epipharynx is retractile due to the muscle from the outer clypeal wall and close the food channel at the base of the proboscis mechanically and by the lacinial lobes of the maxillae.

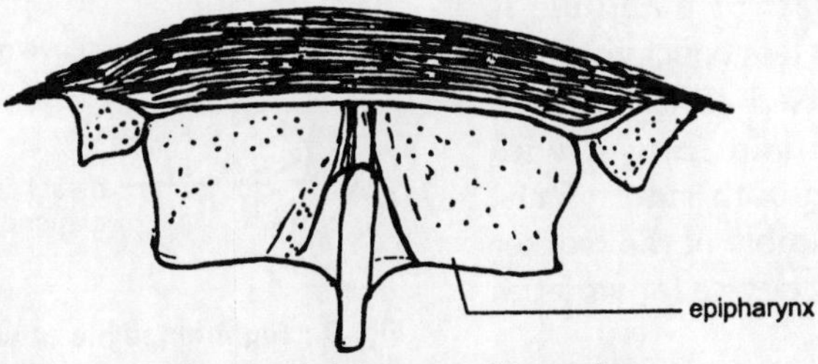

Fig. 8 : Epipharynx of honeybee

The Mandibles

In all three castes of the honeybee, the mandibles are vary with the shape and relative size.

Drone Mandible

The mandible of drone is smaller than that of queen and worker of the honeybee. Mandibular hairs are long and numerous. The distal part of drone mandible is narrow. It is with small apical tooth and has mesal depression from which a groove leads upto the base.

Queen Mandible

Female mandibles are the plumose variety. Outer surface is clothed with more numerous and longer hairs than the worker bees. In female, mandibular hairs are unbranched. The length of queen mandible is similar in length as that of the worker but, wider at base and bilobed at distal end. Outer lobe is strong. Groove is absent on the orifice of the gland.

Worker Mandible (Fig. 9)

The mandible of worker is wide at the base, narrow at middle region and widened again at distal region with a flattened expansion and rounded margin at the end. It is articulated anteriorly on the lateral angle of the clypeus and posteriorly on the gena. It also has narrow and flexible membranous attachment to the lower margin of the gena. It is provided with a pair of abductor, a pair of adductor and adductor apodeme (Fig. 9). The distal inner part of mandible is more or less concave and is transversed obliquely by a channel and fringed with hairs on both sides. Thus, the mandible of the worker is handy device for varieties of work.

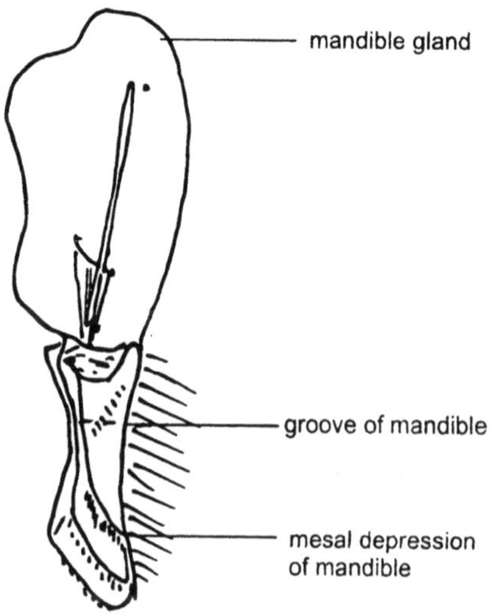

Fig. 9 : Right mandible of worker

Mandibular Glands

A large mandibular gland is present on the mandible of worker. It is sac like and extended upward to the level of the antennal bases. The gland in drone is very small but, it is larger in queen. The gland has an epithelium of secretory cells lined with a thin cuticular intima. The function of mandible gland in bees is not known clearly.

The Proboscis (Fig. 4)

The proboscis of honeybee is complex structure made by the fusion of maxillae and labium. It is a functional unit madeup from maxillae and labium in the form of tube through which liquids are drawn up to the mouth by the action of cibarial pump. Proboscis is kept folded when it is not in use. However, the median elongated, straight structure with long hairs at the end point is located as labrum and the lateral elongated, straight structures are the maxillae. From the membranous floor of the proboscis fossa, these three parts are suspended on the back of the head and further united as the lorum by 'V' shaped sclerite.

Special Features

Maxillary palpi are very small. Galea is with its concave inner surface forming a roof over the glossa and fits lengthwise against the labial palps forming food channel. Labial palps are long which form food channel with galeae. Glossae units forming a channel called alaglossa which can reach deep into nectary of blossoms. Nectors are sucked up more rapidly through glossa.

The larval feeding organ is greatly simplified in form and structure and adopted to need of special feeding habit. The labium has a broad base from underside of the head and fused laterally with simple and slender maxillae. The labium has free lobe from distal region to which hypopharynx is united dorsally.

The Maxillae (Fig. 4)

The maxilla shows following parts : a long basal part coxa, a segmented palpus which represents the telopodite and two coxal endites, the galea and lacinia arising mesad of the base of the palpus. The coxal base is divided into proximal cardo and distal stipes by an elbow point. The maxilla of the adult bee shows all parts and most of the muscles found in general insect. However, shapes and proportional sizes are different in bees.

A suspensory basal rod is articulated on the margin of the proboscis fossa and large boat shaped stipes at its posterior end. Stipes at its distal end provides laterally a very small palpus which is two segmented and the continuing part is tapering blade which is long and thin and pointed is the galea found in worker bee. The galea has prominent midrib and it is narrow at the base transversed throughout its length. The small triangular basal plate is subgalea found on outer side of the stipes. The galea and stipes contain hairs of various sizes. The maxilla has the association of usual types of muscles as noted in generalized maxilla. However, the maxilla is more strongly protracted by three large muscles inserted on its posterior part. The maxillae has no retractor muscles as in other insects. Thus, proboscis is entirely dependent on retractor muscles of the labium for retraction. The lacinia is prominent lobe which arise on the end of the stipes mesad of the subgalea. The lacial lobe is called as segelhalter.

The Labium (Fig. 4)

The labium of honeybee consists following parts :

1. post mentum
2. prementum
3. palpus
4. glossa }
5. paraglossa } lingula

In general, it is belived that the honeybee labium maintains all the parts of generalized labium. The labium of honeybee worker consists a small triangular postmentum and an elongate prementum. From prementum diverging, slender segmented bodies are given out called palpi. Between the bases of palpi arise the lingular lobes, and long median tongue. The tongue is combined glossae and a pair of short paraglossae. In the membranous wall of the proboscis fossa, postmentum and the base of the prementum are located between the bases of maxillae. Postmentum narrow base lies closely in the angle of lorum. The divergent arms of post-mentum are articulated on the distal region of the maxillary cardines. The maxillae and labium are joined together in such a way that the entire proboscis becomes protractile and retractile on the cranial articulations of the cardines. However, the labium has no direct connection with the cranial walls.

The labial palpus has four distinct segments. It has also a small membranous basal lobe which is attached to the prementum. The basal lobe is known as a palpiger. Glossa and paraglossae, the lingular lobes have common base at the end of prementum and have a pair of lingular arms anteriorly and posteriorly by a triangular sublingular plate. The tongue of honeybee is densely hairy. Closely ringed structure and narrow transverse bands of sclerotization having hairs alternate with bare membranous spaces are the important characteristics of tongue. Therefore, the tongue is flexible and contractile. Distally the tongue tapers and ends with a small spoon-shaped lobe called the labellum. The labellum is with smooth and convex surface posteriorly and armed with free margin. Small branched hairs are located on its concave anterior surface. The thin, elongate lobes arising from the common lingular base on the prementum are called paraglossae. Extrinsic muscles from the head, muscles confined to the prementum and the muscles from the basal segments of the palpi is the musculature complex of labium. Labium of the larva is a simple lobe with no palpi or lingular lobes and is united with hypopharynx.

Labial glands

Labial glands are ectodermal in origin which develop from ingrowths of labial segment of the head.

Types of Labial Glands

1. Head gland (or) post cerebral gland.
2. Salivary glands.
3. Thoracic glands.

Classification

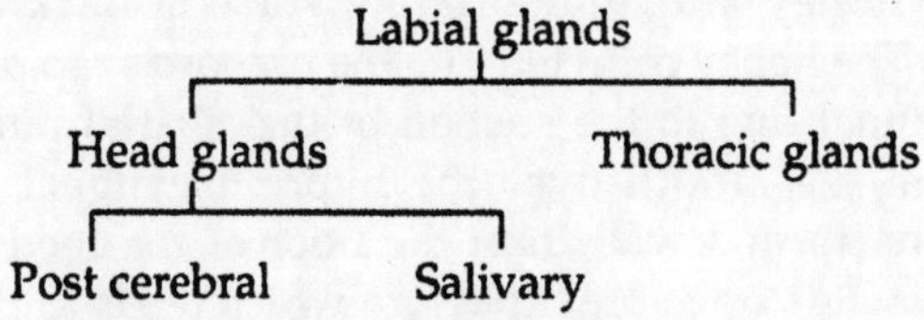

1. *Head glands* (Fig. 20)

The head glands are situated at the posterior walls of the cranium. The head gland of worker consists of a loosely arranged mass of small pear-shaped bodies having separate ducts found uniting irregularly with each other and finally forming a single duct joining the common median duct from the thoracic glands. Head glands are vestigeal or absent in the drone caste of the honeybee but, found well developed in worker and queen.

2. *Thoracic glands*

In a single thoracic gland, a mass of many branched glandular tubules are found opening into short collecting ducts. These ducts further combine to form several major ducts and form a saclike reservior anterior of the gland. A out duct is given from reserviour which joins with the ducts from the glands in the back of the head. The thoracic glands are present in all castes of honeybee as 'salivary glands' of the other insects.

The common duct of the labial glands opens at the base of labium and the secretion is passed within the paraglossae to the back of tongue, to tongue cannal and discharged upon labellum and mixed with food taken from the channel of proboscis. The salivarium function as a salivary ejection pump or salivary syringe. It probably sucks the gland secretion from the duct and forcefully expels it from the distal opening into the space between the paraglossae at the base of the tongue. The muscles act here as compressors for salivarium by flattening the chamber after its expansion by dilator muscles.

Galea and labial palpi comes together around the tongue and form tube through which liquids may be drawn up to the mouth by sucking mechanism of cibarium during functional stage of proboscis. During feeding on honey, a worker bee dips the end of proboscis into the liquid, the free of the tongue is bent back on the surface of honey and immediately starts a back and forth movement. The honey collected into the proboscis rapidly ascends the food cannal due to the suction of the cibarial pump and is poured into the mouth over the hypopharyngeal lobe. The hypopharynx form a wall from the floor of the food cannal of proboscis to wide open oral apurture which leads to the cibarial chamber. During feeding behaviour of bee, the tongue may have different and rapid movements produced by the single pair of muscles attached to the base of glossal rod. In nonfunctional stage of proboscis, its distal parts galeae, labial palpi and the tongue turned back into the posterior side of the head. Folding of proboscis is simple but, anatomical events are much more complex.

The Thorax (Fig. 10)

The thorax of honeybee contain three segments. Anterior segment attached to head is prothorax. The third segment of thorax attached to petiole of abdomen is metathorax and the middle segment is mesothorax. From ventral side, in each segment one pair of legs is given out. Fore legs are attached to prothorax, mid pair to mesothorax and hing legs to metathorax. Two pairs of membranous wings are present in honeybee. Thorax in honeybee and hymenoptera is more complex and difficult for identification of the different parts. Thorax is the mobility centre because, legs and wings are attached to the thorax. The dorsal sclerotized part within the segment is the tergum. Notum (Fig. 10) is combine part of pro-meso

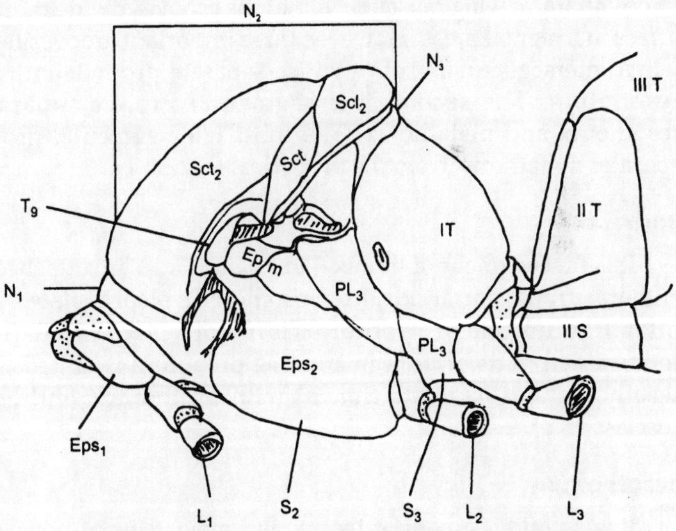

Fig. 10 : Thorax and petiole of abdomen of worker

Epm	–	Epimeron	Sct	–	Scutum (prothorax)
N_1	–	Notum (prothorax)	Scl_2	–	Stutellum
N_2	–	Notum (mesothorax)	Sct_2	–	Scutum (meso)
N_3	–	Notum (metathorax)	I T	–	Tergum propodeal
Eps_1	–	Episternum (prothorax)	II T	–	Tergum (propod.)
Eps_2	–	Episternum (mesothorax)	III T	–	Tergum (Adb.)
L_1	–	Fore leg	II S	–	Sternum
L_2	–	Mid leg	Pl_3	–	Pleural plate
L_3	–	Hind leg			

and metathoracic sclerotization. The ventral sclerotized part within the segment is sternum and lateral sclerotized part is pleurum or pleura which is wing bearing part lies between the wings above and the part between the legs below. The wing bearing notum shows notus sulcus, scuttum and scutellum and also phragma, anterior and posterior notal wing process, post scutellum or post notum. The thoracic pleuron shows following parts– pleural sulcus, episternum, epimeron, basalare, subalare, etc. while, the thoracic sternum shows basisternum (eusternum), furca sternum (sternellum), endosternum and furca, etc.

Specialities of Bee Thorax

Propodeum is present on the thorax as 4th segment. However, it is not true segment of thorax but, it is more sclerotized plate of abdomenal second segment extended over the metathorax. The thorax has maximum solidity for maintaining stress put upon it by mesothoracic wing muscles. Numerous grooves divide thoracic surface in irregular areas. However, three important grooves refers to first intersegmental groove which separate propodeum from mesonotum. The second intersegmental groove separates mesothorax and metathorax. The third intersegmental groove separates metathorax from the propodeal tergum.

Prothorax

It is first segment of thorax. First pair of legs are attached to prothorax from ventral side, from prothoracic pleural plates. The collar like notum of prothorax is wholly separated from pleurosternal parts of its segment. The pronotum is united with mesothorax and have attachment with prothoracic and head muscles.

Mesothorax

It is second segment of thorax and major part of honeybee thorax. Ventrally, a pair of mid legs is attached to mesothorax. The space between fore leg pair and mid leg pair is more than the space between mid and hind pair. From dorsal side, a pair of fore wing is attached to notum.

Metathorax

It is third segment of thorax. Metathorax contain a pair of hind legs and a pair of hind wings on ventral and dorsal side respectively. The wings are membranous and transparent. The coxa of hind leg is articulated laterally on a small condyle at the junction of the ventral pleural margin with propodeum. Propodeum is large sclerotized dorsal plate and a narrow ventral plate. The dorsal plate or tergum is strongly convex and attached to abdominal petiole. Two spiracles are seen on propodeum which are located on its lateral sides.

Thoracic Appendages

The legs and the wings are the thoracic appendages of honeybee.

Legs (Fig. 11)

There are three pairs of legs in bee attached in each segment of thorax – at some what different angles from ventral side. Each leg of honeybee contain coxa, trochanter, femur, tibia, tarsus, pretarsus, ungues and arolium. In addition to walking, the legs of worker are modified for various other purposes. The legs are important tool of honeybee for its various needs. The bee is an artist which can perform for various works within the colony and outside the colony by its legs and mouth parts. The legs are used for clearing antenna as it is a sensory part, and pressing pollens/nectors.

Coxa (Fig. 11)

It is basal part of leg which is attached to the thorax between plura and sternum. As there is slight difference in articulation of legs in coxal axes, the three pairs of legs have their little own different movements. The legs can turn forward and backward. The fore legs have different forward and backward movements because, the coxae hang downward on transverse axes between plural and sternal articulations. The hind legs are attached on narrow posterior part of the thorax, therefore, they are generally directed to backward. However, the mid and hind legs turn outward with forward movement and inward during posterior movement. This is happened due to coxal articulation which is placed obliquely on thorax with their slanting axes. The coxae are provided with remotor and promotor muscles.

The Trochanter (Fig. 11)

It is second joint of the leg which has dicondylic articulation to the coxa at horizontal transverse axis and to the base of femur at dorsal and ventral articulation. The trochanter has only one reductor muscle from femur. It is very short part of the leg.

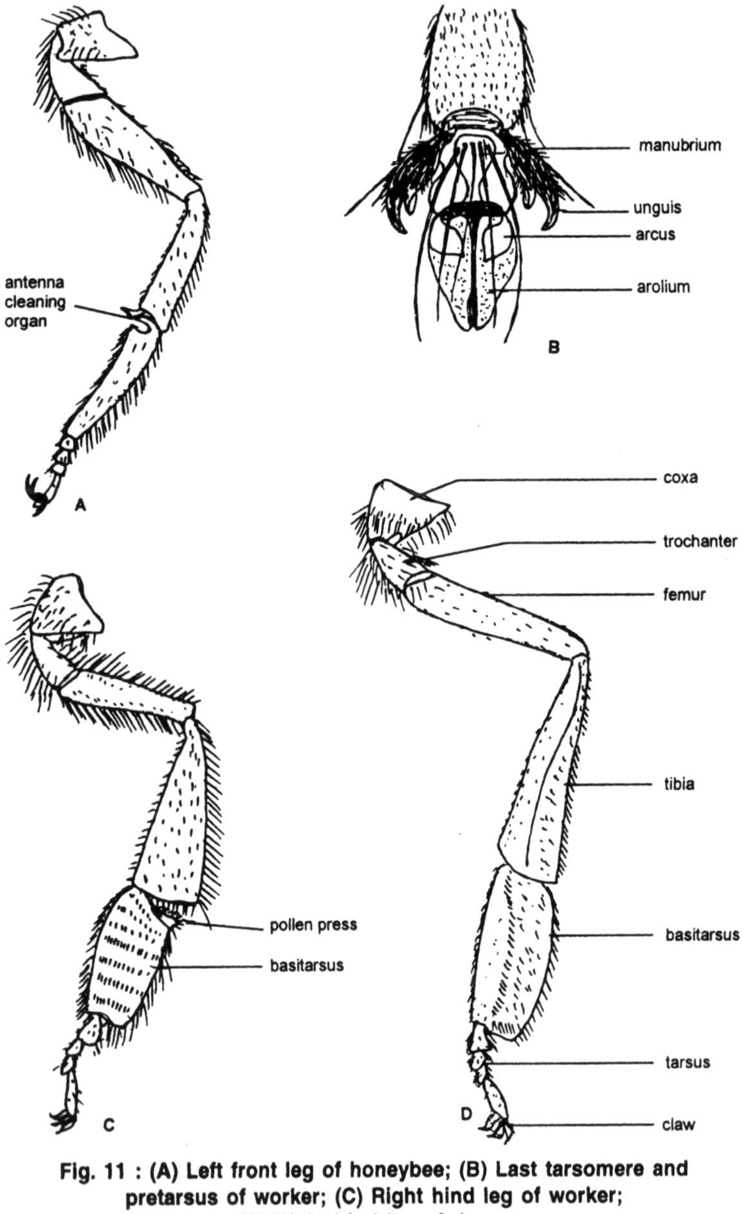

Fig. 11 : (A) Left front leg of honeybee; (B) Last tarsomere and pretarsus of worker; (C) Right hind leg of worker; (D) Right hind leg of drone

The Femur (Fig. 11)

The femur is elongate part of honeybee leg. Basically, it is attached to trochanter and distally to the tibia. Tibial muscles are found throughout the length of femur. Muscular complex includes, dorsal extensor and large ventral flexor. The femur also contain branched flexor muscle of pretarsus and a fine tendon like thread.

The Tibia (Fig. 11)

The tibia in leg has its own speciality in bee. The hind tibia is greatly elongated, flattened and wide distally. While, the fore and mid tibia are slender and shorter than femur. In drone and queen, hind tibia is more slender. The head of all tibiae is more or less bent while, the end is deeply emaginate and broad. In worker, hind tibial outer margin is smooth and more or less concave and boarded by fringe of long incurved hairs thus, forming basket which is refered as "corbicula". In corbicula honeybee worker collect and transport pollen and propolis to hive. The tibiotarsal joints are monocondylic. In all tibiae, between tibia and tarsus there is a single articulation which is median and dorsal. The tibia shows tarsal muscles and a long branch of retractor muscle from pretarsus.

The Tarsus (Fig. 11)

The tarsus is segmented part of the leg. Thus, it is divided into five segments called tarsomeres. Basal joint (tarsomere) is much longer than others. Four segments of tarsus have monocondylic attachment to each others and freely movable. Muscles are absent here for helping their movements. The tendon of the flexor muscle of pretarsus is helpful in transversion and deflection of tarsus. However, in hind leg it is broad and flat. The basal tarsal joint is known as metatarsus. "Antenna cleaning organ" is located at the base of the tarsus. Basitarsus of hind leg is similar in size (large size) in each caste of the honeybee. In worker, posterior surface of basitarsus is broad and armed with nine transverse rows of long spines directed distally with 45 degree thus, forming flat brush. The brush is useful for collection and retension of pollens to be stored in tibial basket. Between basitarsus and tibia of hind leg, there is deep notch in dorsal margin which is converted into "pollen press" in worker.

This is used in transfering pollen from basitarsal brush of one leg to the tibial basket of the other.

The Pretarsus (Fig. 11B)

It is complex and specialized structure. The median part of pretarsus is attached to last segment of tarsus. The pretarsus has a pair of lateral claws. These claws are called as ungues. Claws arise from the base and articulated separately on condyles at the end of tarsus. The arolium is soft apical lobe of pretarsus and the end part of pretarsus. Arolium is adhesive organ and generally turned upward between the ungues. The basal part contain elongate, flask shaped median sclerite which is armed with 5–6 curved spines. Proximal to arolium is broad weekly sclerotized plate, the planta. Proximal to planta is unguitractor which is a large shield shaped plate. Distally the unguitractor is provided with two small auxiliary sclerites. The claw has two long outer spines. In worker and queen claw is similar in shape. In drones it is angularly bent and having long, slender points.

The Antenna Cleaner (Fig. 11A)

The antenna cleaner is present in most Hymenopterans but, it is well developed in honeybees. It is reported in worker, queen and drones. In worker bee, "Antenna cleaner organ" shows following parts :

1. Deep semi circular notch on the ventral surface of basal part of basitarsus.
2. A large, flattened spur arising from the inner angle of the tibial distal end.
3. Closing like a clusp or fibula over the tarsal notch.
4. The fibula is movable, thin and broad appendage. It is narrow at base and sharp at distal point and with strong accessory lobe to its anterior part but, with no muscles.

During the processess of antenna cleaning hoenybee makes appropriate movements of the fore leg so that the tarsal notch is first kept around the base of antennal flagellum and then due to flexion of the tarsus, the antenna is brought against the fibula and held in cleaner. The antenna is then moved upward, and the tarsal comb cleans the sensory outer surface and then the thin accessory febular lobe scrapes the inner surface of antenna.

Collection of Pollen and Propolis

The foraging bees collect pollens and nectors from flowers and from plant buds propolis is collected. The above materials are collected by the worker bee from the field into the basket of hind tibia. Propolis is resinous gum obtained by the bee from plant buds which is used for sealing the cracks and crevices in the hive and strengthening wax for building of comb. When honeybee visit flowers, pollen get settled all over the body of bee. These pollen are collected by bee using pollen collecting apparatus. However, honeybees specifically displace the pollen from anther during foraging. The pollen associated with anterior part of the body are moistened and regurgitated with honey and make them sticky and adherent. The pollen grains are then cleaned with the help of pollen brushes of the front legs. Further, scraped on the broad inner surfaces of basitarsus of mid leg. Mid legs grasp one at a time through basitarsal brushes of hind legs and called forward and transfer the pollen to inner surface of hind basitarsus. The full filled basitarsial brushes then fill the pollen into baskets of tibia. Between the tibia and basitarsus a pollen press (a notch) is located whch play very important role in transfer of pollen. The tarsal lip in worker is widened, beveled outside, upward the tibia and expand as a small lobe.(auricle). The opposing armed notchlip form comb (pecten) or little rake (rastellum). When the basitarsus turns upward at tibiotarsal point, the fringed margin of auricle presses within the semicircle of long curved hairs on the lower end of the floor of pollen basket. Hind legs brought together and moved alternately up and down in the process of pollen presses and filling the pollen basket, due to which, the rastellum of descending legs scrapes pollen in small quantity from basitarsal brush of the other leg. The pollen separated by rastellum rest on the surface of the auricle and get attached to it. Further, due to an upward flexon of basitarsus on tibia, pollen on auricle are pressed up into lower part of the tibial basket. The baskets are loaded by repeatative scraping, first from one basitaral brush and then from the other and operation of the presses. When both pollen baskets are loaded, the bee returns to its colony, goes to the hive cell and store the pollen in a store cell. Honeybee rests her front legs on one edge of the cell, then place the end of her arched abdomen against the other, stretch the hind leg with pollen to the cell and pore the pollen load with basitars of mid legs to the floor of cell and go back for foraging. Other bees then place the pollen properly, from bottom to top in the cell.

For collection of propolis, bee first bites the buds with the help of mandibles and fore legs, then a small piece of gummy exudation is extracted. The particle first kneaded in the jows and then taken on fore leg (any one). By the time, hind leg of the same side is brought forward, particle of propolis is scraped to mid leg by fore leg and immediately pressed into the corbicula of hind leg. This action is repeated on other side. Then honeybee visits the comb, the matter is brought to needed place, waits for some time for another bee in the hive. Then the propolis is used for cementing or repairing the comb at needed place by another bee.

2. The Wings (Fig. 13)

The wings of insects are evolved from integumental folds, the paranotal lobes located on the mesothorax and metathorax. Due to favourable musculature and flexibility of the insect body, insects are unsurpassed flyers. In honeybee wing develop internally, complete development of wings takes place beneath the cuticle of pupa. After emergence from pupa, the adult bee spread its wings, clean them and flew away.

As like other flying insects except diptera, the honeybee also contain two pairs of membranous wings with well developed wing venation. One pair of wings is located on mesothorax and they are called as fore wings. The another pair is situated on metathorax and the wings are called as hind wings. Fore wings and hind wings are different in size, shape and venation. The fore wings are much larger than hind wings. Fore wings and hind wings are coupled with each other by a coupling apparatus. Due to coupling, wings becomes strong from the view point of flight. The honeybee wings are more simplified with respect to shape and venation when compared with other insects.

In honeybee coupling apparatus consists of a row of minute hooks on anterior margin of hind wing. The hooks curve upward and backward and having twisting appearence. The fore wing is provided with a narrow fold on its posterior margin. When the wings expand for flight, the fore wing is drawn over upper surface of the hind wing and hooks are articulated to fold of the front wing. However, at resting stage, and when wings folded over back, there is no articulation between the fore and hind wings. In bees the wings are given out from lateral side of the thorax between the notal and pleural plates on meso and metathorax. Each wing is articulated to the edge of notum and supported with wing process of pleuron. The pleural fulcra lie at short distance laterad to the notal hinges and alternates downward and upward movements of the notum which lead wings up and down movements respectively.

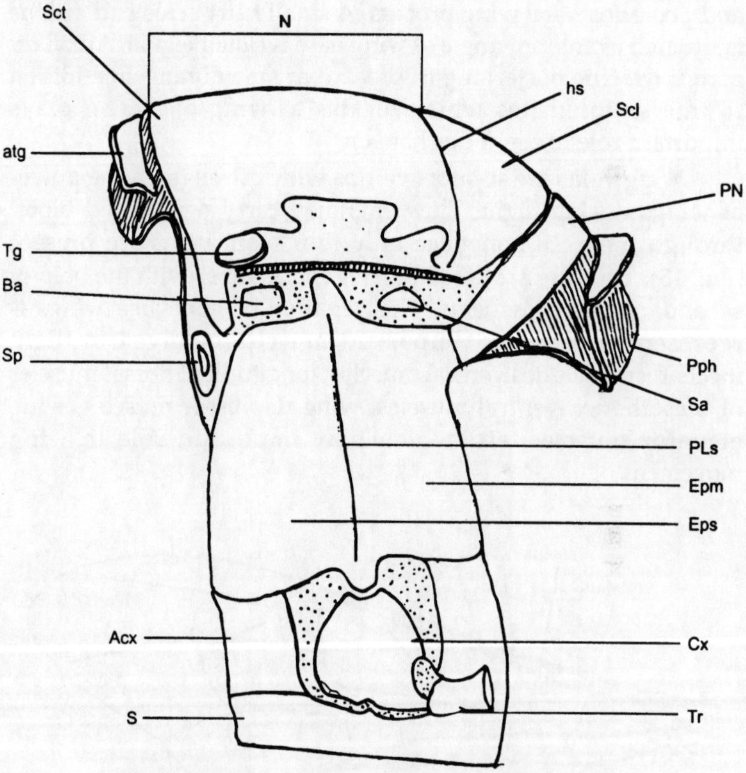

Fig. 12 : Wing bearing thoracic segment structure

atg	–	Acrotergite	Eps	–	Episternum
Tg	–	Tegula	Pls	–	Pleural su'cus
Ba	–	Basalare	Pph	–	Post phragma
Sp	–	Spiracle	Pn	–	Postnotum
Acx	–	Antecoxal plate	Scl	–	Scutellum
S	–	Sternum	Hs	–	External notal sulcus
Tr	–	Trochanter	N	–	Notum
Cx	–	Coxa	Sct	–	Scutum
Epm	–	Epimeron			

Two layered membranous basal part of wing is flexible. The auxiliaries present in basal part of wing are involved in controlling wing movements. In basal part there are three auxiliary sclerites. First auxiliary is attached to anterior wing process of notum (Fig. 12), second auxiliary extends ventrally and settle on pleural wing process and the third auxiliary directly attached to posterior notal wing process which acts as functional flexor sclerite of the wing. In addition, the 4th auxiliary projects from between third auxiliary and posterior notal wing process. A small hairy scale pad or lobe is situated at anterior angle of wing base is called tegula. Auxiliary cord is the true posterior part of auxiliary membrane. In addition to this, epipleurites which locates at wing base also plays important role in flight mechanism.

Wing veins are supportive ribs which strengthen the power of wings during flight. The veins may carry nerves and blood through them. In honeybee only four main veins are present (Fig. 13). They are scattered in irregular manner with the help of secondary and crossveins. The venation of honeybee wings is represented in figure. Important muscles involved in flight mechanism includes ventral muscles, longitudinal dorsal muscles of mesothorax, ventral muscles, wing depressor muscles, wing elevator muscles, etc. which play important role in wing movement.

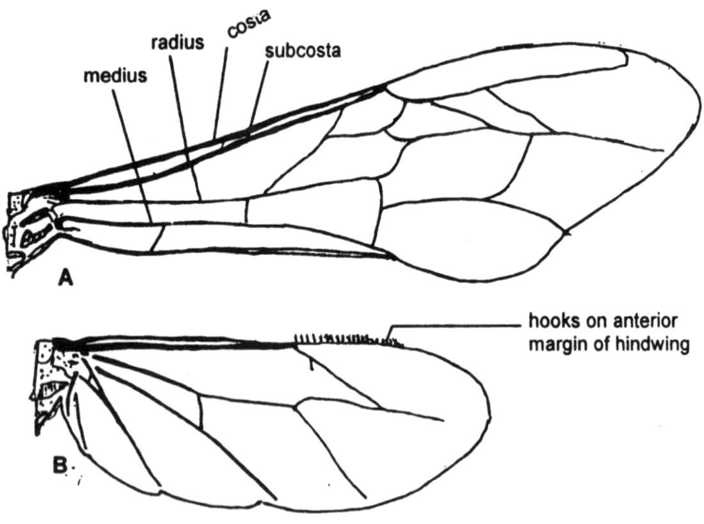

Fig. 13 : (A) Fore wing; (B) Hind wing : honeybee

The Abdomen (Fig. 14)

The abdomen is an important body part of bee because, it includes most parts of the various systems of honeybee. Abdomen contain 10 segments but only 9 are functional. The segments of abdomen are simpler than thorax. The sclerotized part of each segment located on dorsal side is tergum, ventral is sternum and lateral sclerite is latero tergite or laterosternite. In worker bee, abdomen is broad anteriorly and abruptly narrow at the petiole and tapers to the posterior end. The abdomen of drone is rounded at posterior end. However, the abdomen of queen is similar in shape as in worker. There are six prominent segments in abdomen, these are II to VII. Three segments are found concealed within the segment VII. These three segments shows modifications related to sting of the bee. The abdomen of drone ends by tergum of VIII segment and sternum of VIII segment is concealed within the VII and IX segment represents and ends with prominent sternum. The abdomen of honeybee is divisible into three regions namely,

1. Pregenital or visceral region
2. Region of relatively unmodified segments
3. Genital region

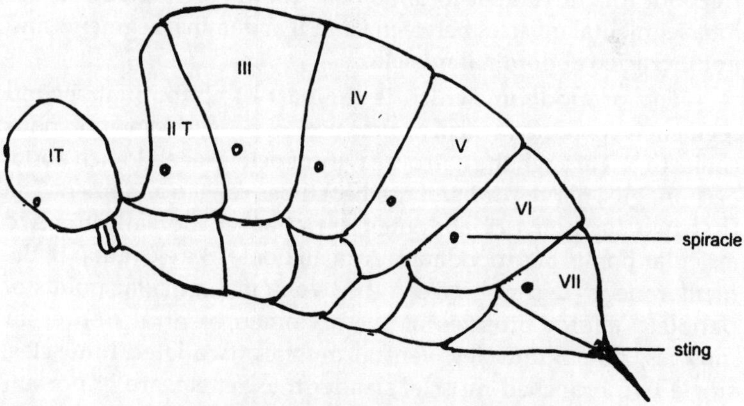

Fig. 14 : Abdomen of worker lateral view

1. Pregenital region

The abdominal segments from III to VI in female and from III to VII in male comes under pregenital region of the bee. The segments of this region are well developed with sternal and tergal plates. Terga overlaps the sterna along the side of abdomen. In the lateral part of terga spiracles could be seen. The tergum is anteriorly supported by internal sub marginal ridge called antecosta. The antecosta provides short apodermal process from each side which project forward into the neighbouring segment. The connecting membranes are also folded (Fig. 15) which allow extensive movements to the sterna. Wax glands are present on the anterior part of sterna from segments IV to VII. In worker, the wax pockets are located in the space between the two sterna.

Fig. 15 : Infold intersegmental membrane

The Petiole

The first two segments of abdomen which attached to metathorax are highly modified, they form petiole of the abdomen. The petiole thus, formed has great importance in giving much freedom for movement to abdomen. The muscles involved are intersegmental muscles between I and II abdominal segments and not thoracico abdominal muscles.

The propodeum which is the part of abdominal second segment is connected with a narrow neck of petiole by membranous conjuctiva. Through the narrow cavity of petiole of abdomen aorta, oesophagus, nerve cord and tracheal tubes runs to abdomen. The first two segments of abdomen are attached dorsally for two articular points of propodeum. A small dome like elevation of the membrane is present between the two dorsal articular points of petiole. Levator muscles of the abdomen, ventral depressor muscles, dorsal muscles, ventral muscles, two lateral muscles, single two branched muscle, slender muscle, etc. are important muscles of this region.

The Genital Region

The genital region is scattered from the segments VII to IX of abdomen. However, these segments are more or less modified and extremely different in both sexes, *i.e.*, male and female.

In drones abdomen ends with strongly decurved tergum of segment VIII and reduced sternum of the same segment and normally concealed with VII segment. VIIth segment of drone has same features and tergal musculature as noted in preceding segments but, abdomen ends by conical seventh segment and from apex sting is found projecting in the case of worker and queen. Sternal and tergal plates cf VII segment are not different from that of preceding segment except free posterior part of sternum. The segments VIII, IX, X and sting are consumed by segment VII. In drones sternum IX is well developed and crecent plate but, tergum is represented by a small pair of sclerites which is concealed under tergum VIII.

The external parts of the male genital organ (Fig. 36, 37) includes:

1. A small plate on IX sternum
2. A pair of large free lobes between two IXth sterna.
3. A large apurture (phallotreme) between the median lobes from which complex intromittent organ (penis) is given out at the time of mating. A short, tubular anus bearing proctiger is diversing from beneath the 8th tergum.

In 8th segment on lateral plates, a last pair of spiracles is seen. It is complete annulus segment with numerous walls. Spiracle bearing plates are very closely associated with sting called spiracle plates of the sting. On dorsal surface of it intertergal muscle is located, from ventral angle inter segmental sternotergal muscle is present, from ventral margin a long muscle is dilated. The 9th segment provides main support to the sting. The 9th segment is with two lateral tergal plates, forming sting apparatus (Fig. 19) called quadrate plates, below this are oblong plates. The anus bearing 10th segment of the female is similar to male segment. However, in retracted position, it lies in the sting chamber, beneath the 7th tergum. Below the base of sting there is opening of female reproductive system.

Abdominal Glands

The honeybee has following types of abdominal glands :

1. The wax gland (Fig. 16)
2. The scent gland (Fig. 17, 18)

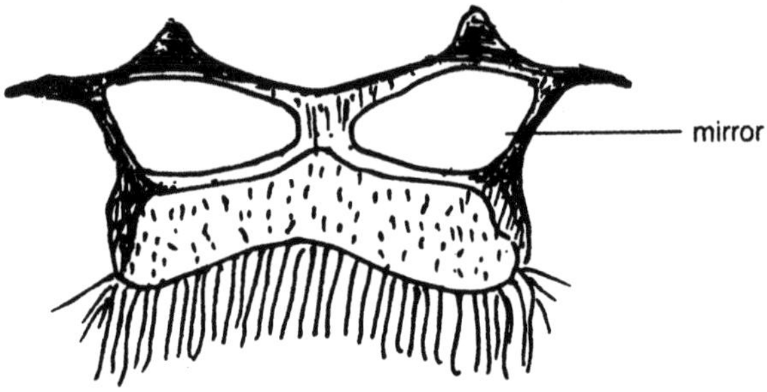

Fig. 16 : Wax gland of worker

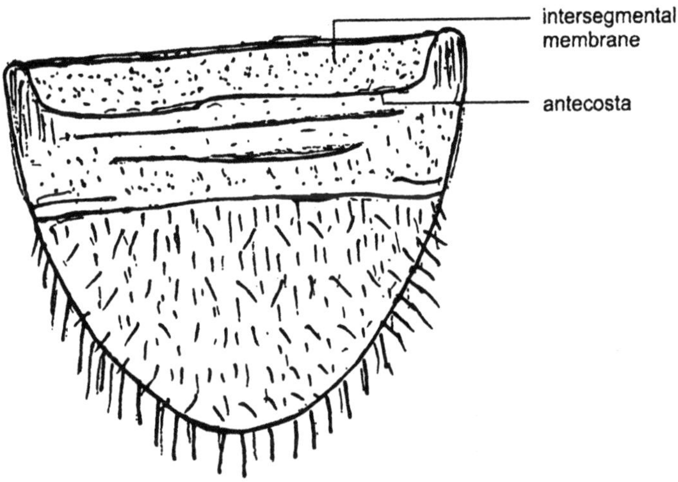

Fig. 17 : Abdominal tergum VII of worker

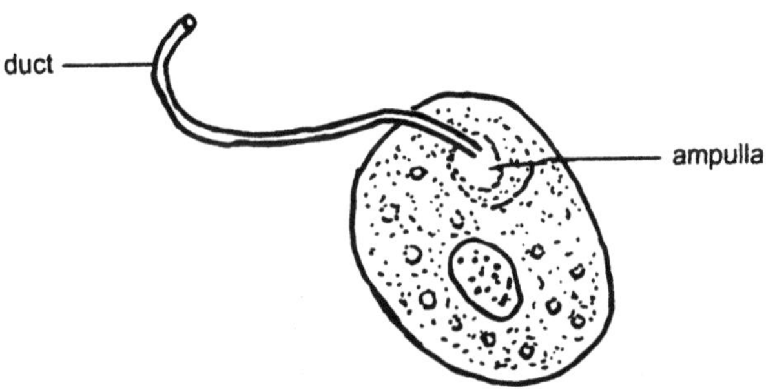

Fig. 18 : A single gland cell and its duct

3. Other epidermal glands, and
4. Sting gland.

***Wax gland* (Fig. 16)**

In worker bee, on the anterior part of sterna of abdominal segments IV to VII wax glands are present. Each sternum of above segments have two large, oval polished surfaces which are framed by dark marginal bands. The polished two surfaces are separated by a broad median space. These surfaces are called as wax mirrors or wax plates. Above them there are four pairs of wax glands. The wax glands are having epidermis over each gland as large cellular mass composed of fat cells and oenocytes. The thickness of gland is vary with the situation, time and age of the bee. In young workers epidermis are with ordinary flat epithelium but, in active bee, cells are found elongated vertically and shows clear space between them. In old bees, most of the cells shrink, in over winter worker epidermis becomes simple sheet of nucleated plasma. Actively secreting wax gland cells have longitudinally fibrillated appearence and contain numerous basophile granules and supplied with tracheae and tracheoles penetrating into the cells and intercellular spaces.

The brushes on the basitarsus of the hind legs are used by bee in removing the wax scales from wax gland/wax pockets. The hind legs are pushed back with the tarsal brush from under surface of the abdomen one at a time and the spines on the brush collect the scale from the wax pocket. The scale generally sticks to the brush. One scale is taken from one side and other from adjacent side or two or three may be removed successively. The leg then bent to forward and wax is taken off by the mandibles and fore legs, then masticated in jaws and taken to the comb where construction of comb part is needed.

***The Scent Gland* (Fig. 17, 18)**

Abdominal scent gland is present beneath the intersegmental membrane which is weakly sclerotized, basal part of tergum VII. The scent gland lies below the anterior half of the pale area of tergum VII. The scent gland is with numerous ductles which open into the basal canal. In worker the scent gland appear as cluster of large glandular cells. They are situated beneath the canal of tergum and anterior elevated half area behind the canal of tergum.

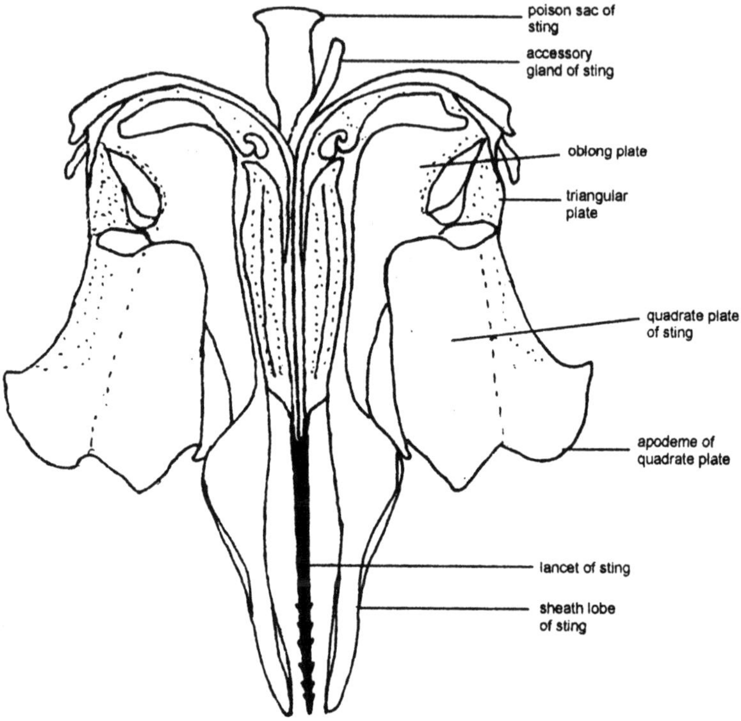

Fig. 19 : Sting of worker bee.

The cells of the gland have a delicate and hair like individual ducts arising from ampula of cell and opens on the floor of the canal through cuticle. The ducts appear as great number of minute filaments attached to the inner surface of cuticle. Their number varies from 500 to 600. In drone, scent gland is not present. According to Jacobs scent gland is also absent in queen. But, according to Mc Indoo it is present in queen of the honeybee.

4.
Anatomy of Honeybee

The Digestive System (Fig. 20)

The alimentary canal of honeybee consists of pharynx, oesophagus, crop, proventriculus, ventriculus, malpighian tubules, proctodaeum ; intestine, rectum and anus. Mesenteron of bee is endodermal in origin and stomodaeum and proctodaeum are ectodermal in origin.

Alimentary canal of bee is not a single structure, it under goes various changes in life stages such as embryo, post embryo and adult. First adapted for serving the embryo shutup in an eggshell yolk stored in egg is food for embryo. The alimentary canal structure is further adapted for the need of larva resting in comb. In the pupal stage the entire food tract undergoes a radial transformation and doing different functions. The embryonic develpment of bee in the egg requires 3 days. During this period the embryo subsists on the original supply of yolk. In adult stage, larva requires different food hence stomodaeum opens into mesenteron. Ontogenic development is an adaptation of old processes to new conditions. The larva of bee has a complete alimentary canal from the time of hatching.

In larva, alimentary canal is simple which contain following parts :

1. Short, slender stomodaeum
2. Mesenteron which is long, cylindrical and extended through the length of body
3. Proctodaeum
4. Anus
5. Large malpighian tubules, are given out between proctodaeum and ventriculus. The ventriculus represents

simple epithelium of cubital cells. Peritrophic membrane lines the epithelium. During the pupal stage alimentary canal is greatly enlarged attaining the length which is sufficient for adult bee.

Alimentary Canal of Adult (Fig. 20)

The Stomodaeum

It contains pharynx, oesophagus, crop and proventriculus.

The Pharynx

The pharynx is single sac like body appear in the head. Its walls and lumen are continuous along with the preoral cibarium part.

The Oesophagus

The oesophagus is slender, simple tube which passes from pharyngeal upper narrow end to abdomen through thorax. In the abdomen it expands as large crop which is sac like body. The walls of oesophagus are lined with thick cuticular intima. Numerous circular folds of wall are located, therefore, expanding the body area of oesophagus internally. A strong muscular layer is located on outside which is composed of external circular fibers and inner longitudinal fibers.

The Crop

In drone and queen the crop is more slender than worker. In worker, the crop is just enlargement of the oesophagus in the abdomen. The crop walls have similar structure as noted in oesophagus. It is commonly known as "honey stomach". However, physiologically, it is not a stomach at all. The crop of honeybee is just like a large balloon shaped bag which is filled by nectars. Thus, it is used by the bee for filling the nectars in it and transporting the nectars to the hive for conversion of nectars into honey. When it is fully filled, it looks like large balloon and when empty it becomes small flabby pouch.

The Proventriculus

The proventriculus in queen and worker is more or less similar but, queen crop is slender and more symmetrical. The proventriculus is a short and narrow part of alimentary canal

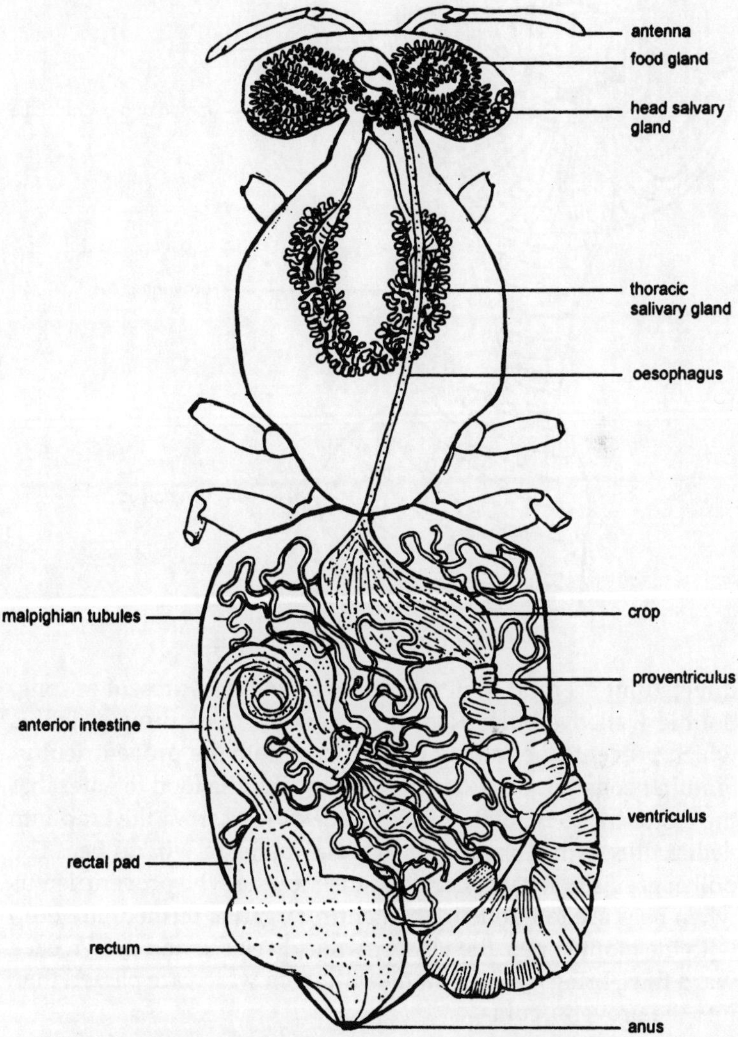

Fig. 20 : Digestive system of honeybee (worker)

which lies between crop and ventriculus. Its anterior part invaginates into the crop and posterior part is attached to ventriculus, thus, it appears as a neck of crop. The invaginated part of proventriculus has opening into the crop by four triangular lips closing by "x" shaped opening (Fig. 21). This opening is called as mouth of the proventriculus. The lip walls are provided with dense cuticular intima. The lips are armed by the group of spines

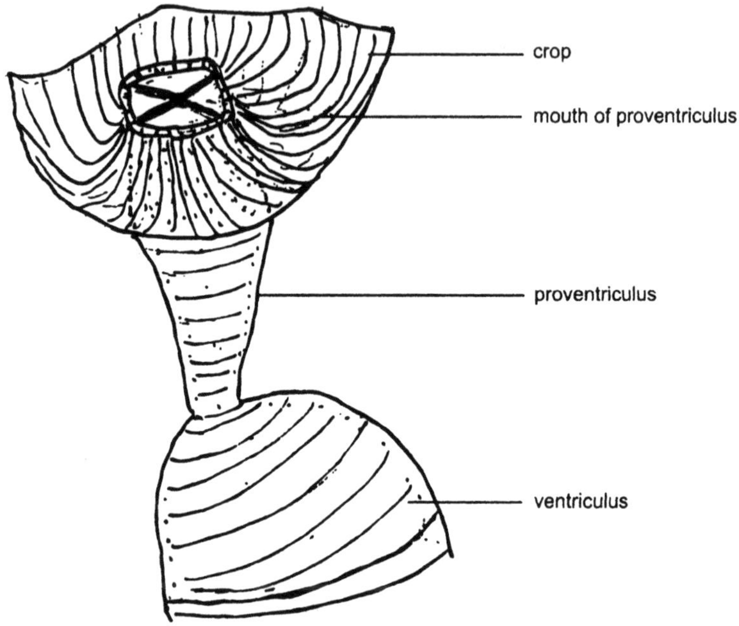

Fig. 21 : Mouth of proventriculus

directed into central lumen. Stomodael valve is present as long, double walled tubular fold of posterior end of proventriculus which prevents the forcing of food back into the proventriculus. Simultaneously, it provides the passage, for the food to enter into the stomach. To regulate the entrance of food from the crop into ventriculus is the function of proventriculus in worker bee. The pollen grains first get collected at the mouth of the proventriculus when bees are fed pollen. Then entire organ is pushed into crop with the mouth open, then lips get quickly closed and drawn back ward thus, brings pollen grains with them which are passed then into the proventriculus.

The Ventriculus (Mesenteron) (Mid gut)

The ventriculus is largest part of alimentary canal of honeybee. It is thick and cylindrical tube of 'U' shaped loop and acts as functional stomach of the bee. The proventricular surface is brownish and has numerous transverse constrictions which form deep internal folds on it. The epithelial wall of ventriculus starts from the base of the outer lamella of stomodael valve and reach to the end of ectodermal part of canal and the beginning of the endodermal stomach. The epithelium has major transverse folds which are formed by infoldings of the cell layer, basement membrane and circular muscles.

The ventriculus has secretary function, the epithelium of ventriculus also serves for excretion, particularly of calsium.

The honeybee infested by *Nosema apis* discharges watery fluid causing dysentry in the bee. In such cases, spores accumulate on large scale in the walls of ventriculus. The secretary cells of ventriculus epithelium are reduced resulting in disorganized work of cells and filling the spores in proventriculus and its walls. The spores are also found in malpighian tubules which affect rectal activities, cause deleterious effects, generate watery fluid in rectum and cause dysentry in bees.

The Peritrophic Membrane

There are different opinions regarding the origin and manner of formation of peritrophic membrane. The food in ventriculus is covered by several thin, irregular coverings. These cylindrical coverings are called as peritrophic membranes. They are formed from whole length of the ventricular epithelium in honeybee. According to Weil (1935) the peritrophic membranes of bee arise from the rhabdorium. While, Von Dehn (1933) says that the peritrophic membrane is a direct product of the epithelial cells and chemically it is chitinous in nature.

The proctodaeum **(Hind gut)**

The proctodaeum has two main regions namely, the anterior intestine and the posterior intestine is called as rectum. The anterior intestine is slender tube anteriorly and opens into ventriculus by widened mouth. It is looped towards the posterior end and opens into rectal pad of rectum (Fig. 22). The epithelium of ventriculus form the fold over the mouth of the intestine. The anterior intestine is truely small intestine due to small tube. The epithelial wall of anterior intestine provides six longitudinal folds which are persistent in most of its length. Outer layer is of circular muscle fibers. However, longitudinal muscles are absent in anterior intestine. The intestine where it joins the ventriculus is called as pylorus. In the pylorus lumen, all malpighian tubules are open just behind the ventriculus. A thick fold of the intestinal wall forms the pyloric spinctor or pyloric valve (Fig. 23) on the mouth of malpighian tubules. The posterior end of anterior intestine tapers and join the rectum. There is crowding of six folds of inner wall at the joining point which forms puckered exit orifice. A large, thin walled sac which lies in the posterior part of abdomen is called rectum. The rectum joins to the slender anterior intestine by its immediate narrow part and posteriorly it becomes tapering for joining into the anus which is situated in greatly reduced 10th segment of abdomen.

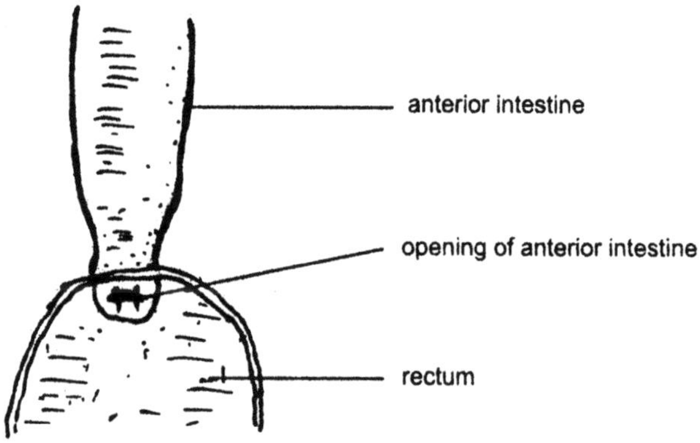

Fig. 22 : Opening of Interior Intestine

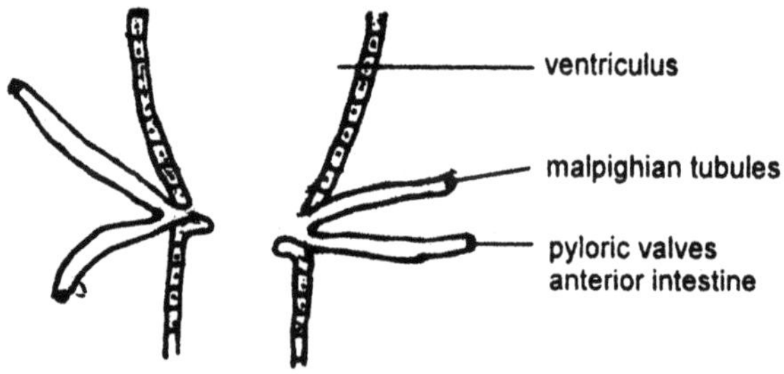

Fig. 23 : Pyloric valves and opening of malpighian tubules

The epithelium of rectum is a layer of thin, nucleated protoplasmic layer of thin cuticular intima and having numerous small folds. Outer sheath is of circular muscle fibers and external layer of smaller, widely separated longitudinal fibers. The epithelial folds and intima provides rectal sac which hold an immense accumulation of faecal matter. It also hold the excretion of malpighian tubules. In honeybee, rectum is expanded enormously and occupies entire space available in the abdomen when the pressure of intestinal wastes increases. Honeybee never excrete in hive and in prolonged cold periods, the faecal matter retained in the rectum for longer period. This is the speciality of rectum of honeybee. The rectum has six long thickenings of epithelial wall called rectal pads/rectal papillae. Water is absorbed in rectum probably in rectal pads. The simple thickenings of rectal epithelium are called as rectal pads which are ectodermal in origin (Dobrovsky, 1951).

The Malpighian tubes (Fig. 23)

The malpighian tubules are long, whitish, convoluted, thread like tubes. They are coiled and more than 100 in number in bees. They open into the anterior end of the intestine. Malpighian tubules are anatomically the part of alimentary canal but they have independent function of excretion. Thus, malpighian tubules are excretary organs. The walls of the tubule consists of single layer of cells with striated border and outer basement membrane. The excretary products are in the form of small crystals of nitrogenous and other substances such as uretes, phosphates, calcium oxalate, calcium carbonate, leucin, etc. The excretory matter first accumulates as a secretion in the inner ends of tubule cells and excreted through the striated border in free globules. As regards to the origin of malpighian tubules there are different opinions. The tubules are believed to be the outgrowths of pyloric region of proctodaeum and are ectodermal in origin.

Physiology of Digestion

Salivary glands release the saliva which contain inverten. The juice taken by the bee from flowers react with inverten and forms invert sugar during the transformation from field to the hive. In oesophagus excess moisture is absorbed from the food material. Before reaching the honey stomach, this matter (juice of flowers) is converted into the honey. However, the honey can not enter into the ventriculus because of the "x" shaped valve which is present in proventriculus, that permits only appropriate quantity of honey needed by the bee. In ventriculus digestive enzymes are secreted which helps to digest the food either liquid or solid. In ventriculus, the nutrients from food goes into the blood for keeping bee in good condition. The poisonous food or rancid food is purified by malpighian tubules of the bee.

The pollen in the stomach of honeybee is completely digested without mechanical or break down of the walls. However, some pollen grains pass unchanged. Here digestion may takes place perhaps through the microplular membranes. Pollen starch is easily digested unless broken.

The pharyngeal glands (kind of salivary gland) produce "royal jelly" at early days of adult life. The "royal jelly" is fed to queen larva and young brood. However, the glands needs pollen for full activation during adult age of 3-6 days. These glands secrete proteolytic and acidic secretion. Further, pharyngeal glands start secreting enzymes such as amylase and invertase in foraging bee of 3 week old and the glands become most active in one month old bees. They also secrete an oxidizing enzyme which form acid (gluconic acid) from glucose. The resulting acidity preserves the honey properly. The labial glands of posterior division (cephalic gland) produce clear oil for utilizing the wax, and saliva used for moistening the material for chewing. The thoracic glands produce watery secretion of pH 6.3-7 which is used in building the hive. In foraging bees thoracic glands are also responsible for producing invertase.

The Circulatory System

Majority of insects are having air tubes for the transporting oxygen. The blood of the insects acts as ash collector. It collects waste products of metabolism from the tissues and carries to the special excretary organs. In insects, the circulatory system is scattered throughout the body cavity which is not occupied by the tissues. Through pulsating dorsal vessel, the blood is kept in motion. The dorsal vessel extends from about posterior end of abdomen to the head beneath the brain.

The circulatory system is mesodermal in origin. In the lateral or upper part of the mesoderm bands, the coelomic cavities are only small cleavage spaces. The dorsal vessel, a dorsal diaphragm, a ventral diaphragm and pulsating organs in thorax and head are the important circulatory organs of adult honeybee.

The Dorsal Vessel

The dorsal vessel extends from the head to the posterior half of the VIth abdominal segment of adults. But, in the larval forms it extends upto the ninth abdominal segment. The heart (Fig. 24) which lies in abdomen increases in width to the posterior end of abdomen. It contains 5 pairs of lateral ostia. The ostia are lateral openings of the heart. The first pair of ostia is located in abdominal segment II and last pair is on VIth segment.

The oblique vertical slits to the lateral walls of the heart are the ostia. Ostia opens into the lumen between long flaps. These flaps extends inwardly and anteriorly in the lumen. The heart is semicircular and thin walled tube. It contains semicircular muscle fibers dorsally and ventrally in its wall. The heart has the origin from paired rows of cardioblast. The heart wall is with sarcolema of muscles and without endothelial lining. The heart is connected from the dorsal body wall by fine connective tissue strands and from the diaphragm by fibrils. It is also provided with muscles on its lateral and dorsal walls. Anteriorly, the musculature of heart is weaker but beyond convoluted part of aorta it is extremely faint and with flat nucleated cells. The heart is with sufficient tracheal complex which is supplied by segmental branches of the abdominal lateral large air sacs. The tracheal branches of heart units and open into the dorsal sinus.

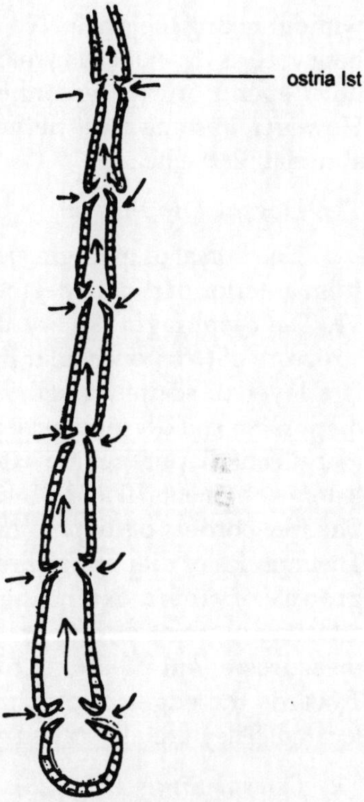

Fig. 24 : Diagramatic horizontal section of heart showing ostia and blood flow

Due to the contraction of heart, the blood is driven forward but prevent backward flow. The costal lips open and allow free entrance to blood during dilation of the heart. The costal lips divide the heart into chambers and the lips also act as valvular structure within the heart. Generally, the heart beats from behind forward by successive waves of contraction which pass anteriorly. The heart continues its rhythmic pulsations by diaphragm muscles or without taking their help. No reversal direction of beat is there in honeybee as it is noticed in many other insects. The heart beat does not depend on nerve connections from any source in bees. However, in some cases nerve cells are the automatic centers for stimulating the heart.

The Dorsal Diaphragm

The dorsal diaphragm extends from upper abdominal cavity from anterior part of III segment to the posterior part of segment VI. The diaphragm is very thin and transparent and complex structure of two noncellular limiting membranes. Between these, is a layer of scattered cells and are muscle fibers. The heart, diaphragm and dorsal trancheal sinus is represented in figure. The margins of diaphragm are extended into points and are attached to tergal segments III and VI and antecostal apodemes at their base. The free borders of diaphragm forms deep semicircular notches. The muscles of diaphragm are in five pairs which are fan shaped groups of fibers extending toward the heart from lateral attachment point of diaphragm on the tergal plate. The fibers measures about 10-20 micron in width and are slender one. Towards the edges of diaphragm, diaphragm cells are loosely located. They are also called as paracardial cells.

The Dorsal Sinus (Fig. 25)

The dorsal sinus is the large air sac which contain dorsal tracheae. These tracheae extends to heart and diaphragm for aeration. The mass which covers the heart contain fat cells likely, these cells are also located against the dorsal wall of sinus. The dorsal sinus is part of the body cavity which is above the dorsal diaphragm. With abdominal visceral cavity it is in free communication. The dorsal sinus opens into tracheal air sac.

The Ventral Diaphragm

The ventral diaphragm begins from thorax where it has attachment with composite endosternum of mesothroax and metathorax. It extends in abdomen via propodeum and petiole and found suspended by lateral attachment point on the anterior marginal apodems of sterna. In worker and queen, the diaphragm becomes narrow and forked into two prongs. The prongs are attached to the spiracular plates of segment VIII anteriorly. This diaphragm has no cellular elements and supporting membranes.

It has musculature slender than fibers of the dorsal diaphragm radiating from their attachment of sterna. In contraction of abdomen, the diaphragm is thrown into wide transverse folds. It pulsate as strong wave like passing posteriorly.

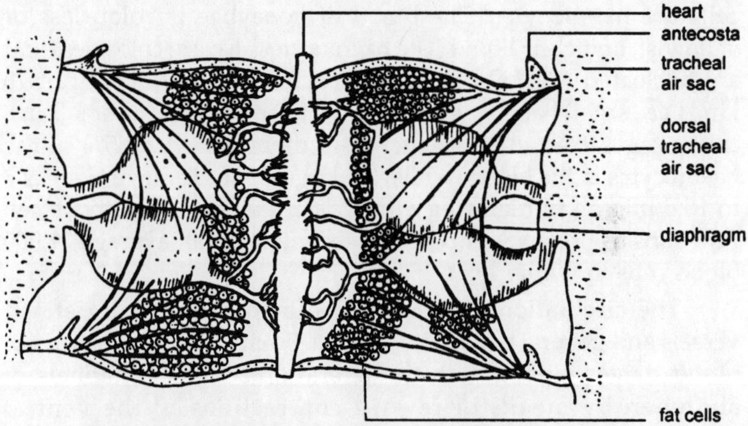

Fig. 25 : Dorsal sinus and organs in abdominal segment

The Ventral Sinus

The ventral sinus lies below the ventral diaphragm which communicates with the abdominal cavity above it by lateral openings along the borders of diaphragm. The abdominal nerve cord cross the upper part of sinus where fat cells are located as a ventral layer. The ventral commissure with sac like swelling crosses the sinus from one side to other between the fat cells and nerve cord.

Accessory Pulsating Organ

Accessory pulsating organ is reported in bees very first in 1911 by Janet. The pulsating organ helps to maintain the flow of blood in antenna and wings in bees. It is located between the bases of the antennae as a thin walled vesicle. The vesicle expanded by the contraction of pharyngeal muscles and blood drawn into it. The elastic walls of the vesicle then contract and drive blood through the lateral vessels into the antennae. A muscular membrane stretched beneath the air sac in the hump of scutellum which draw the blood up from the bases of wings to the space around the air sac and then run into the abdomen for discharging.

The blood returns along the sides of the thorax from the head then enters to wing bases.

The Blood and its Circulation

The blood of insects is refered as haemolymph and the blood cells are haemocytes. The blood of honeybee is colourless or yellowish and clear liquid. The haemocytes also are colourless and are nucleated cells. On the basis of size, haemocytes are of two kinds *i.e.*, small haemocytes and large haemocytes. The small size cells ranges from 8-11.36 micron in diameter with 85% of the haemocytes in the blood. While, the large size cells range from 12.7 to 15.6 micron in diameter with 15% of the cells. The above cell functions as phagocytes. The blood contain on an average 21,000 blood cells in winter bees.

The circulation of blood is confined to the well defined vessels and general body cavity in bee. From the ventral sinus, the blood driven up through the lateral and viceral channels of abdomen by means of reward contractions of the ventral diaphragm. Due to the contraction of dorsal diaphragm, dorsal sinus expands and the blood is drawn into the dorsal sinus through the openings along its sides. Within the sinus, the blood is pressed ahead, along the sides of the heart, due to the forward pulsations of the diaphragm and then it enters the heart by way of lateral ostia. Then it get pumped anteriorly through the aorta and finally reach the cavity of the head. From the head, it turns back through the thorax and again enters the ventral sinus of the abdomen. The blood circulation takes place through larger spaces of body cavity. The blood also circulated to the tissues of the body and to the appendages by accessary pulsating organs.

The Fat Body, Urete Cells and Oenocytes

1. The Fat Body

The fat body is loosely aggregated masses of soft tissue. The cells store fat, retain glycogen which is consumed during metamorphosis and in the pupal stage. The elaborate protein substances appearing in cytoplasm as Albuminoid granules, in blood are used as food for developing tissues. Fat body cells (Fig. 26 a,b,c) are derived from lateral part of mesoderm. The fat bodies of adult worker consists of thin layers of cells spread on the body wall of the abdomen specially in the dorsal and ventral blood sinuses. They are also present on each side of the abdomen as a band of fat tissue. In a foraging worker, the cells of fat body contain sufficient amount of fat. While, in young summer bees the fat content of cells is very high showing the cells swollen with large globules. In such cases the nucleus of cells is distorted and may divide.

2. *The Urete Cells*

The urete cells are not present in adult bees but, they are present in larva and pupa which are sparsely distributed among the cells of the fat body. These cells are smaller than fat cells, in the cytoplasm, minute refractile crystals are present and inclusions which are uric acid salts. These cells store nitrogenous excretory matter.

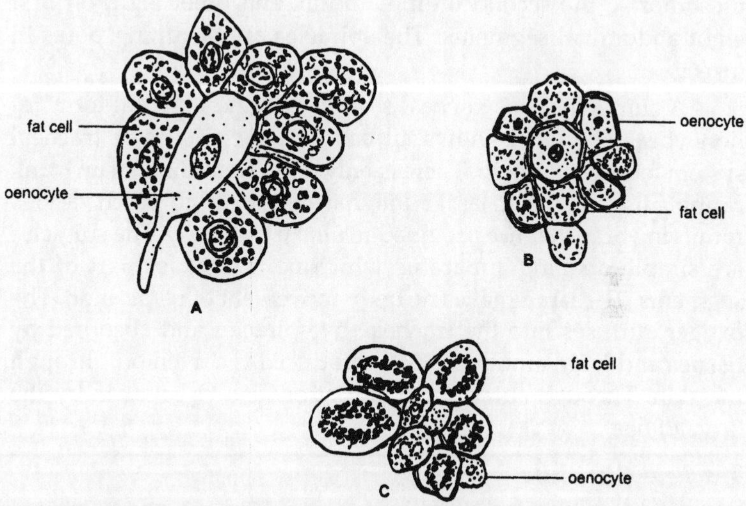

Fig. 26 : (a) Oenocytes and fat cells (showing glycogen); (b) fat cells and oenocytes of over wintering bee; (c) fat cells and oenocytes of foraging worker in spring

3. *The Oenocytes* (Fig. 26 a,b,c)

The oenocytes are ectodermal in origin. They are reported in larva and pupa of honeybee. The oenocytes remain persist during the larval and pupal stage of honeybee. However, they are destroyed and replaced by imaginal oenocytes newly formed from the epidermis of abdomen. The oenocytes are irregularly oval, unattached cells found embeded in the fat body in young bee larva. These cells are much smaller than the fat cells. In matured larva many oenocytes tapers to one end, some cells are with long neck. Oenocytes become large spherical cells in early pupa of bees. These cells are much smaller and without outer clear zone. The function of the oenocytes is unclear. But, some workers believe the secretory function. Thse cells form and conserve deposits of wax and may have function related to production of wax in worker bees.

The Respiratory System

In honeybee there is tracheal respiration. Nelson (1915) described the embryonic development of tracheal in honeybee. Tracheal rudiments appear in an embryo in about 45 hours as a row of ten pits or spiracles. The first pair of spiracle is located on mesothorax, the second on metathorax and other eight on first eight abdominal segments. The spiracles are breathing pores in insects.

A simple type of tracheal system is found in larval forms of honeybee. The larva shows all basic characteristics of tracheal system found in embryo. There is only one longitudinal main trunk running from thorax to last abdominal segment from which various ramifying branches are provided to all parts of body. The spiracles are simple opening apparatus which lie in anterior part of the segments. The larva makes no body movement of respiration. The oxygen diffuses into the tracheae by spiracles and absorbed by tissues and large amount of CO_2 produced is driven out through the skin. Through cutaneous respiration also O_2 is taken by the larva of bee.

Tracheal Structure

The tracheal tubules are the ingrowths of body wall. The trachea contains following parts: The epidermal layer of thin epithelium of flat cells and a delicate cuticular intima. The tracheal intima is with spiral thickenings called taenidea. The trachea divides into fine branches and redivides with still smaller tubules called tracheoles. The tracheoles have contact with tissue cells. In tracheole, liquid absorbed oxygen is present which concludes that O_2 is carried by tracheole through liquid to the decreased O_2 pressure area.

Adult Tracheal System (Fig. 27)

Adult tracheal system is illustrated in Fig 27. It is more complex than larval form. Elaborate air sacs are developed in adult bee. The longitudinal trunk has expansions in the form of air sacs. Most of tracheae are without well developed taenidia and appear as very distensible, multitude of silvery, opaque vessels when filled with air. They have numerous smaller branches in thorax and leg parts and they appear every where on various organs and tissues in body cavity.

From the first pair of spiracles two large tubular lateral trunks arise in prothorax. They further extend through neck into the head after bifurcation giving sub-branches to prothorax and mesothorax. The tracheal branches are given to the first legs and to the large wing muscles in mesothorax and further to the posterior ventral thoracic air sac. The lateral trunks unites at the base of the neck as anterior ventral sac (Fig. 27). The anterior ventral sac has two lobes which are extended in prothoracic part as posterior extensions. The lateral trunk is connected with number of sacs in head region. It connects with air sacs lying at the base of compound eye, another is at the base of mandible and one more dorsal sac lies on the upper part of the face covering the top of the brain.

In metathorax a pair of large dorsal sacs are located at the sides of the propodeum. The dorsal sacs further connected with propodeal spiracles by the short tracheae. A median transverse sac is located above the dorsal sacs. The MTC is narrow which occupies the cavity of turgid mesoscutellum. A pair of lateral sacs is given out by the posterior ventral thoracic sac and tracheae to hind legs and middle legs. In the metathorax various tracheae and air sacs unites forming two large tubes which extends to abdomen by petiole.

Two very large lateral air sacs extend through the first five abdominal segments. Two lateral tracheal trunks joins to these sacs. The huge lateral sacs becomes narrow posteriorly and broad toward anteriorly. Two slender dorsal sacs are also located in the petiole region and above the huge lateral air sacs. These air sacs have constrictions on each segment for specifying the segmental regions. They provide tracheal branches to each segments and terminal segments of the abdomen. The tracheal branches are connected to 7 pairs of spiracles in each lateral side and they have

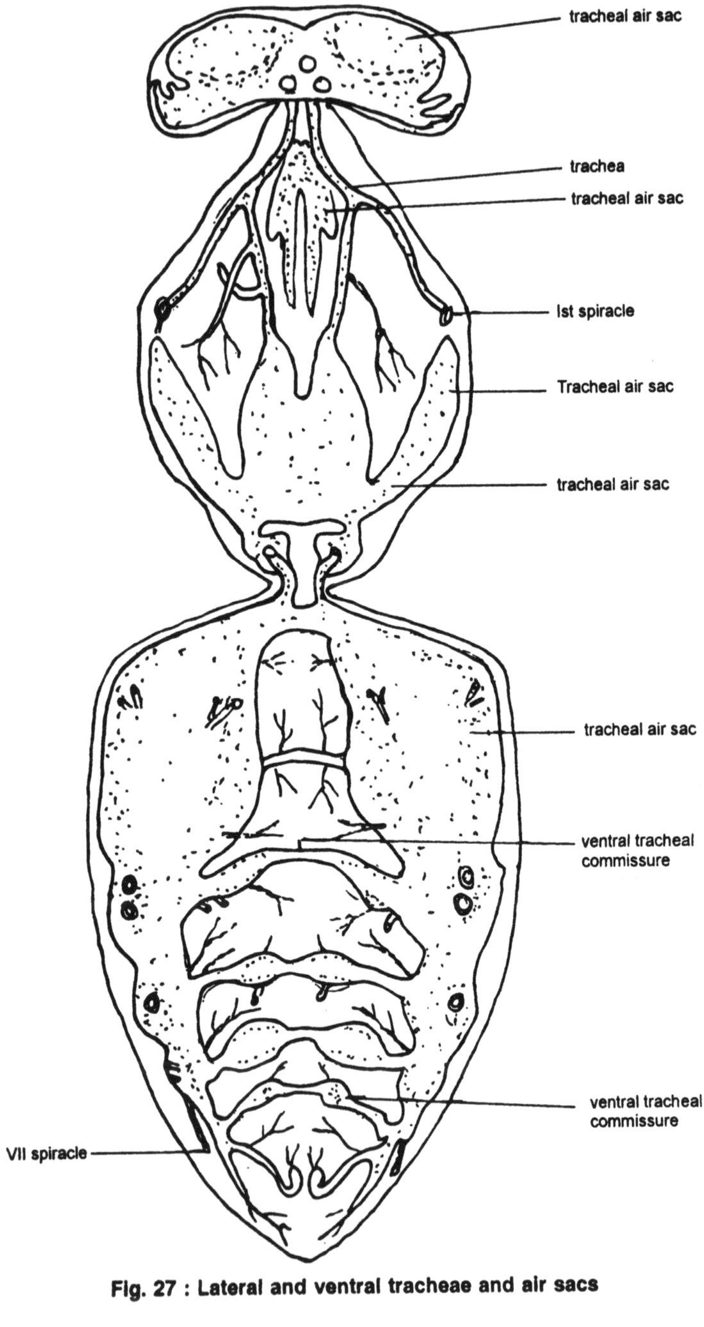

Fig. 27 : Lateral and ventral tracheae and air sacs

six transverse ventral commissures. Out of which 4 have their own small air sacs. Dorsally lateral sacs provide segmental tracheae. Two branched tracheae are given out from the abdominal segments III to VI, which enter the dorsal sinus, heart and surrounding tissues. The thin walled air sacs have no chitinous component and their function might be to give buoyancy to bee in flight. The air sacs serve as storage of air for reducing the weight of bee. However, it is believed that air in sac cannot serve the respiratory needs of bee.

The Spiracles

The spiracles are breathing apurtures in bee. The spiracle is also called as stigmata. Two types of spiracles are seen in bees. Type I are located on thorax and Type II on abdomen. The first pair of spiracle is situated on each side of the thorax below the anterior angle of mesothoracic wing. However, it is wholly concealed under the flat lobe of the pronotum. The spiracles have oval shape and measures about 0.14 mm in length. It is provided with long slender muscle on mesothorax helping for its opening and closing. The second thoracic spiracle is situated in deep membranous fold between mesepimeron metapleuron. It is small sized. The largest spiracles are located on propodeum of the bee on thorax which is first abdominal pair of spiracles. They are surrounded by an elevated cuticular rim. The other abdominal spiracles are situated in first seven segments on the sides of the tergal plates. These spiracles have narrow external openings about 0.06 mm in length. Closing muscles and long dilator muscles help spiracle for opening and closing.

Respiratory Movements

In honeybee, the respiratory movements includes a rapid lengthening and contraction of the abdomen and a slight dorsoventral expansion and compression of the abdomen. This is possible to the bee because, it is provided with muscles both useful for extending and contracting abdomen and for its dorsoventral expansion and compression. The first spiracles are structurally adapted for egress of air rather than its entrance. Bailey (1954) says that air flow through the spiracle has the relation with the activity of the bee and the amount of metabolic carbon dioxide produced. Due to pumping movements of abdomen, the inactive bee inhale and exhale by first thoracic spiracles. However, in active bee, producing high CO_2 tension, inhalation takes place by first spiracles and also by abdominal spiracles and the air is exhaled by way of propodeal spiracles. Under low tension little respiration may be through abdominal spiracles.

The Nervous System

The nervous system has ectodermal origin. The function of nervous system is to co-ordinate all activities of the insect both voluntary and involuntary. The nervous system is highly developed system in honeybees which consists central, visceral and peripheral system.

The elements of nerve system comprises brain, ganglia, nervecord, nerves, neurons/nerve cells. Neurons having cell body or neurocyte and the nerve fiber or axon arising from cell body. Dendrons or dendrites are nutritive roots of the neurocytes. The branch of the axon is collateral. Collateral further subdivide and redivides forming arborizations. The neurons are of three types.

1. Sensory
2. Motor and
3. Association

Synapse is an interlacing of terminal arborizations between the sensory nerve and collateral of motor nerve.

The Larval Nervous System

The bee larva includes following parts as component of nervous system.

The brain, subesophageal ganglion, ventral nerve cord, 3 ganglia in thorax and 8 ganglia in the abdomen. All above compounds are connected by paired interganglionic connectives. The suboesophageal ganglion provides the nerves to mandibles, labium, and maxillae, etc.

The Nervous System of Adult Bee (Fig. 28)

In the adult bee same pattern of nervous system is seen with some modification. The brain is highly developed in adult bee due to well developed compound eyes, simple eyes and the antennae. There are only seven ganglia in the ventral nerve cord of adult bee. There is fusion of four ganglia in the thorax. Likely, two ganglia of abdomen are also fused.

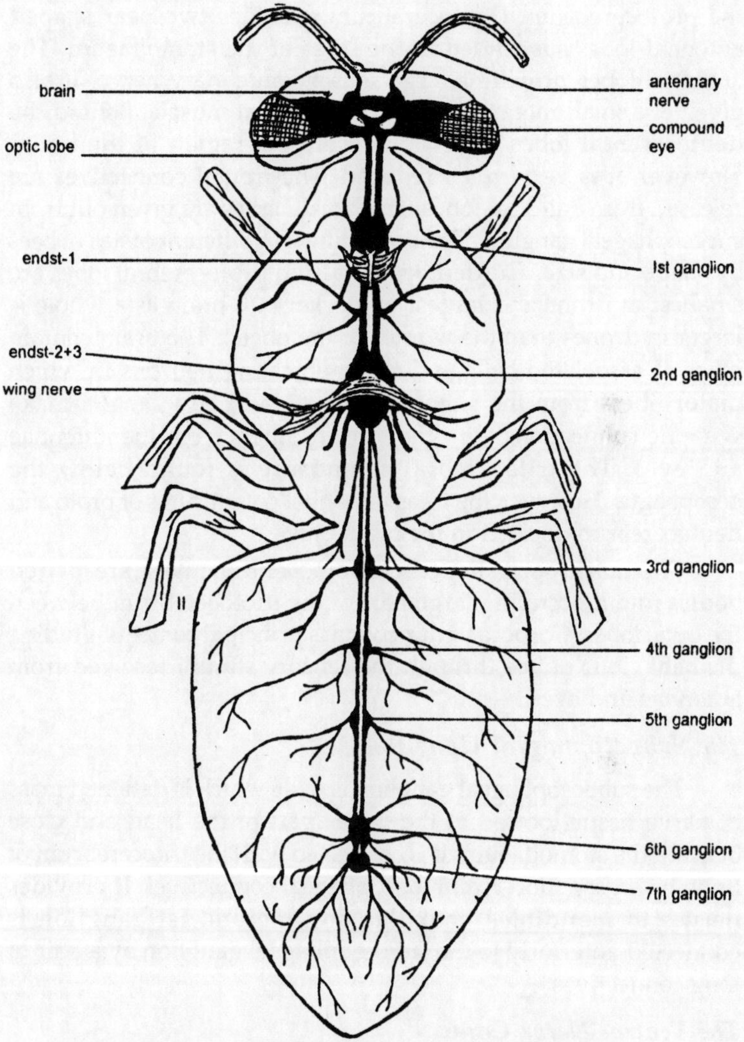

Fig. 28 : Nervous system of honeybee (worker)

The Brain

The brain of adult bee differs from larva as the brain parts are clearly separated and shape is also different. The brain has three regions *viz.*, protocerebrum, deutocerebrum and tritocerebrum. The dorsal protocerebral lobes are broadly joined to each other. The optic lobes have very narrow stalk for joining the protocerebrum. Deutocerebrum provides two pear shaped antennal lobes suspended at the sides of the stomodaeum. The antennal lobes provide a large sensory antennary nerve. It also give very small motar nerve to the antennal muscle. Behind the deutocerebral lobes there is tritocerebral region of the brain. However, it is very much reduced. The frontal connectives are released from antennal lobes. The labial nerves are given out from subesophageal ganglion. The brain differs in different castes of bees by shape and size. The deutocerebral and protocerebral lobes are smallest in drone and largest in worker but, brain as a whole is larger in drones than the worker or the queen. The brain contain mass of associated cells and communicating neurons in which motor fibers from the suboesophageal and body ganglia make synaptic connections with the sensory fibers from the antennae and eyes. The tritocerebral commisure is found below the oesophagus. However, the transverse fiber commisures of proto and deutocerebrum located in the brain only.

The mashroom shaped corpora pendunculata are paired bodies found occuping large part of the protocerebrum between the optic lobes. Corpora pedunculata is principal center of guiding the behaviour of bee through the sensory stimuli received from antennae and eyes.

The Suboesophageal Ganglion

The suboesophageal ganglion is wide which is flattened mass of nerve tissue located at the lower part of the head and close beneath the stomodaeum. It is connected with the tritocerebrum of brain by a very short circumoesophageal connectives. It provides nerves to mandibles, maxillae and labium. Ist ganglion is connected anteriorly to the suboesophageal ganglion by a pair of long connectives.

The Ventral Nerve Cord

It is a pair of longitudinal connectives which runs close to sterna of the thorax and abdomen. It connects suboesophageal ganglion, thoracic ganglia and abdominal ganglia by longitudinal connectives. The nerve cord consists seven ganglia with their connectives. In prothorax, Ist ganglion is located which provides nerves to the first pair of legs. The second ganglion is situated on the combine endosterna of meta and mesothorax. The second ganglion provides the nerves to propodeum, first abdominal segment behind petiole. It also provides the nerves to mid and hind legs, bases of fore wings and hind wings, and to the muscles of meso and metathorax. This is composite ganglion formed by the fusion of probable ganglia of 4 segments as in larva. Third ganglion and fourth ganglion are located in II and III segments of abdomen and provides nerves to 3rd and 4th segments. The 5th ganglion is situated in 5th abdominal segment and the 6th ganglion is composite ganglion which is located on 6th segment but provides nerves to both segments, VI and VII. The last ganglion is situated on anterior part of VII segment of female which becomes representative of segments VIII, IX and X by providing the nerves to them.

Endocrine Organs

Neurosecretary Cells

Weyer (1935) first reported neurosecretary cells in the brain of honeybee. Neurosecretary cells are glandular nerve cells which contain granules of secretion. They are best developed in worker, less in the queen and least in the drone. These cells have secretary activity.

Carpora cardiaca

Carpora cardiaca are little developed in larval forms of bees. In adult bee carpora cardiaca are small paired bodies lying on the dorsal side of oesophagus. According to Hanan (1955) carpora cardiaca are united above aorta and connected to each other by dorsal commissure. These are also secretary organs. However, very little is known about their function in adult bee.

Carpora allata

Carpora allata in adult bee lies against the sides of the oesophagus and close to carpora cardiaca from which small nerves are given out. Carpora allata produce hormone which maintain the juvenile status. Carpora allata increase in size as duble during the larval period. In pupa it decrease in size followed by small increase while in young worker adult it is lost again. In midsummer or overwintering workers it increases to 7 times or more to young adult, in queen increase 5 times and in drones only

3 times. Carpora allata has chief function as to stimulate secretion in the accessory six glands of the male and production of yolk in the ovaries of female. Carpora allata hormone must be regulatory hormone of the general metabolism in bee.

Thoracic Glands

It may probably serve some important function in the juvenile life of the insects.

The Sense Organs

The elements of sense organs such as sensory cells and nerves and central nervous system are ectodermal in origin. Basically, sense organs in insects are of two types:

1. *Exteroceptors* (Respond to environmental stimuli)
2. *Propioceptors* (Respond to stimuli arising in the tissues).

Honeybee has following types of sense organs.

1. Sensilla Trichodea
2. Sensilla Basiconica
3. Sensilla Coeloconica and Ampullacea
4. Sensilla Campaniformia
5. Sensilla Placodea
6. Sensilla Scolopophora
7. The Organ of Johnston
8. The Ocelli
9. The compound eye.

1. Sensilla Trichodea (Fig. 29)

In this case the cuticular part of sensillum is a small hair or seta, present on a membrane in the setal socket over a sublying cavity in the cuticle. The elements of sensilla trichodea are shown in figure. In honeybee trichoid sensilla are numerously present on the mouth parts and antennae. They respond to movements of the seta due to contact with some external bodies. Thus, they are organs of touch.

2. Sensilla Basiconia (Fig. 30)

In this case the external cuticular part has the form of a cone or peg in the form of shortened seta and associated sensory cells.

In bee, peg organs are plenty on the mouth parts and antennae. They occur mostly on the distal ends and last 8 segments of flagellum. The antennae are the principle source of olfactory sense in bee.

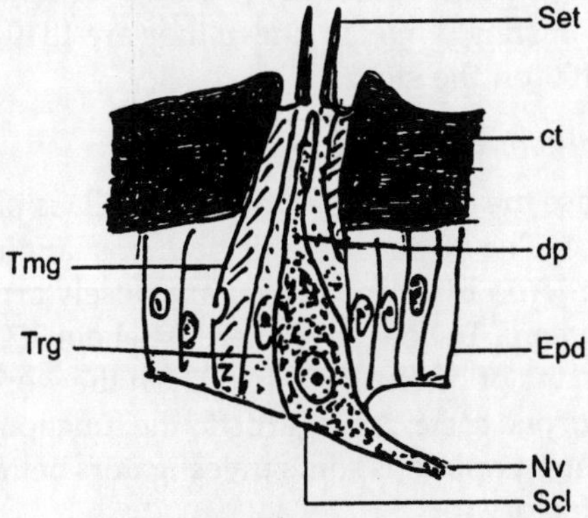

Fig. 29 : Sensilla trichodeum

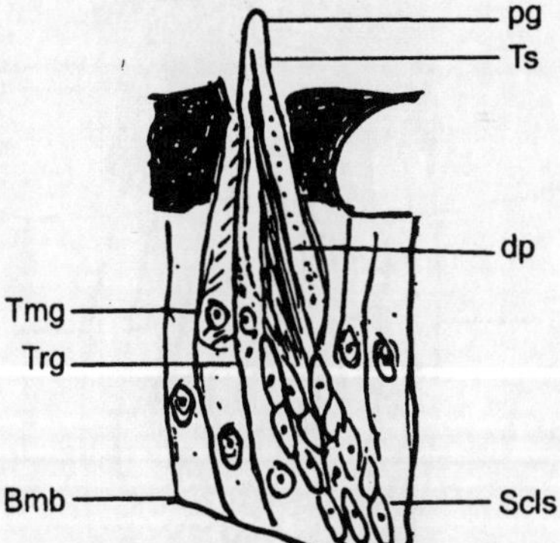

Fig. 30 : Sensillum basiconicum

3. *Sensilla coeloconica and Ampullacea*

In this case the peg is sunken into a pit or a flask shaped cavity of cuticle. These sensilla are found on the antennae of bees.

4. *Sensilla Campaniformia*

It is a conical or dome like body embeded in the surrounding cuticle in such a way that the distal process of the sense cell is inserted like bell clapper. This sensilla occur in small groups on appendages as binucleated cell speciality. In bee the campaniform organs are widely scattered on mouth parts, bases of antennae, wings and sting. In drones, there are 1998 organs at wing bases and 606 on the legs; in worker 15110 on the wings and 658 on the legs and 100 on sting while, in queen they are 1310 on wings, 450 on legs and 100 on the sting.

5. *Sensilla placodea* (Fig. 31)

In this case the external part of the sensilla is plate flush with cuticle surface. The plates are elliptical in shape and measures about 12-14 micron in diameter and are closely arranged among the hairs of antenna. In honeybee, there are about 3000 plate organs on single antenna of queen, 3600 to 6000 in the worker and about 30,000 in the drone caste. As regards to the function of this organ, there are different opinions some investigators believes that these organs are olfactory.

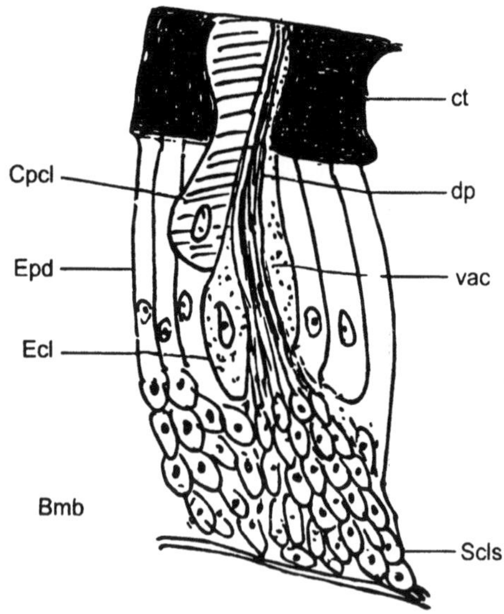

Fig. 31 : Plate organ

6. *Sensilla Scolopophora* (Fig. 32)

A typical sensilla consists of three consecutive cells. The outer cap cell as plate organ, the intermediate cell or enclosing cell, equivalent to the trichogen and the basal sense cell with a long slender distal process. They are present in the head and in the legs. Debaisieux (1938) reported 4 organs in each leg of bee, one in femur, two in tibia and one in tarsus.

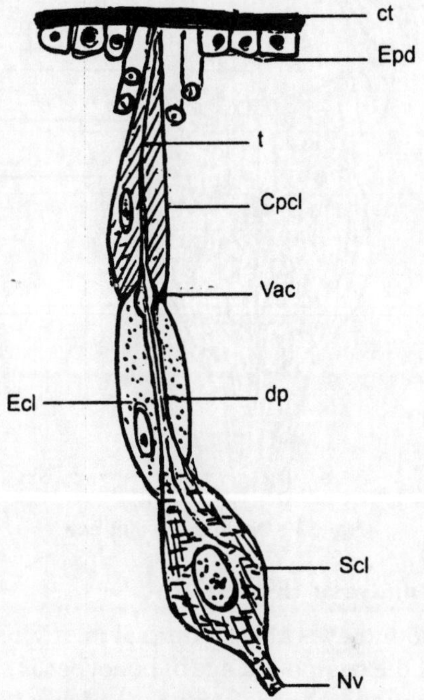

Fig. 32 : Scolopophorus organ

7. *The Organ of Johnston*

It is a mass of small sense cells with long, slender distal processes attached in the cuticular membrane at the base of the flagellum. In bees it needs more details regarding this organ.

8. *The Ocelli* (Fig. 33)

The ocelli are small bodies in the form of simple eyes. They are located anteriorly on the top of the head in triangle in queen and worker bee. In case of drone they are present lower on the face. Externally they are with thick, biconvex lens. The ocelli are just thicking of the head cuticle. The structure of ocelli is shown in Fig. 33. It responds only to certain range of light. The lens unable to throw an image on the retina, may be due to its focal point which is much beyond the outer end of the retinal cells.

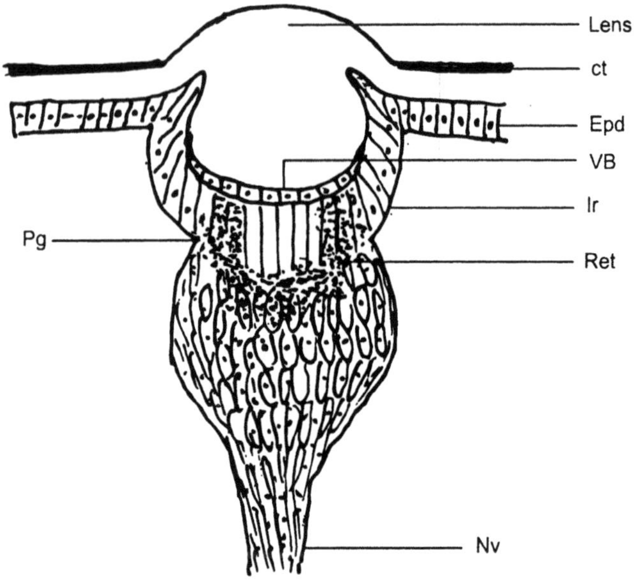

Fig. 33 : Ocellus of adult bee

9. The Compound Eye (Fig. 35)

The compound eyes are ectodermal in origin. The structure and function of the compound eye of honeybee is similar to other insects. The compound eye is composed of ommatidia (Fig. 34). Ommatidia are covered by cuticular cornea and lie close together in the eye. Each ommatidia is separate unit and optically isolated by pigment cells. The unit consists of a distal dioptric part, the cornea with crystaline cone and proximal receptive part, the retinula. The retinula consists of eight retinal cells which are elongated one and contains pigments and extend further into post retinal axon fibre. In a single eye of bee the number of ommatidia ranged from 4000 to 5000 in worker, 3000 to 4000 in queen and 7000 to 8000 or probably more in the drone. In young bees the surface of compound eye is covered with long unbranched hairs of the facets. The structure of compound eye is shown in Fig. 35. The optic lobes of compound eye are ectodermal in origin. The lobe contain a mass of nerve fibers and association nerve cells and shows three distinct internal synaptic regions lamina ganglionaris (outer most), the medulla externa (the next) and the medulla interna (the third).

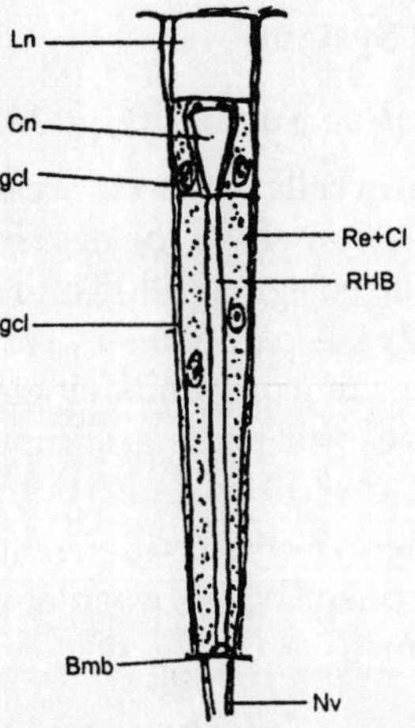

Fig. 34 : Ommatidia of bee

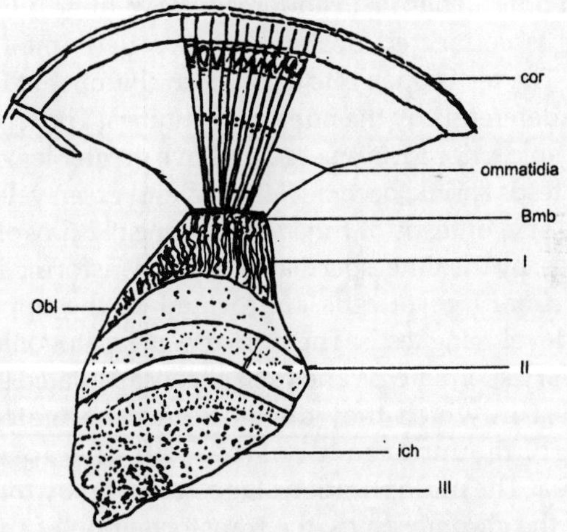

Fig. 35 : Compound eye of bee

Bmb	–	Basement membrane	Pgcl	–	Pigment cell
Nv	–	Nerve	Retcl	–	Retical cell
Opl	–	Optic lobe	Cor	–	Cornea
RHB	–	Rhabdom	ich	–	Inner chiasma of obtic lobe
Ln	–	Lens	I	–	Lamina ganglionaris of optic lobe
Cn	–	Crystaline cone of compound eye	II	–	Media externa
Cgcl	–	Corneagenous cells	III	–	Media interna

The Reproductive System

The Male Reproductive System (Fig. 36)

The primary germ cells of insects are derived from cleavage cells in the early development of the egg. In bee the somatic cells of blastoderm of fertilized egg develop into a female adult and the germ cells become eggs. A pair of testis, vasadeferentia, vesiculae seminales, paired huge mucus glands, single ejaculatory duct, and large intromittent organ or penis are important organs of male reproductive system.

From the mesoderm testis, vasadeferentia, seminal vesicle and mucus glands are formed. While, ejaculatory duct and penis are formed from ectoderm.

The Testis

The testis of matured drone are yellowish, flat, small and triangular bodies. They lie in abdomen above huge mucus glands. Each testis is connected posteriorly to coiled small tube vasa deferentia. The testis open into a common chamber at the anterior end of vasadeferentia by the numerous tubules. The testis in drone of age 4 day old prior to its emergence have greater length but after emergence testis shrink, become flattened and green yellow in body colour. The size of testis of functional drone (12 day old) reduces in such cases, by the time, spermatozoa get transformed to seminal vesicles. Primary germ cells are located in the upper ends of tubules of developing testis. The primary spermatogonium divides into secondary spermatogia and then get encapsulated by thin wall forming syst in which they develop into spermatocytes. The spermatogonia undergo divisions till the formation of spermatozoa. The multiplication stage, followed by the reduction division of the chromosomes, the transformation of spermatozoa developed in round cells and finally the spermatophores move to genital duct from follicle duct. The matured spermatozoa is having the long thread and a head at one end containing the nucleus.

The Vasadeferentia

The vasadeferentia are paired tubues attached anteriorly to the testis and posteriorly to the mucus gland. The vasadeferentia has three distinct parts. The anterior most is short, coiled and small diameter tube. The middle sausage-shaped portion is an large, long and thick called vesicula seminalis. The vesicula seminalis is posteriorly attached to a very large mucus gland by a very short looping tube. In the tubular part of vasadeferens cubital cells are present and in the vesicula wall, long and narrow cells are present. The epithelium is covered with an inner layer of circular muscle fibers and outer longitudinal fibers and membranous tunic lies externally to it. In vesiculae, spermatozoa, coming from follicle tubules are stored where their heads are embeded in the epithelium with their tails filling the axile lumen. Spermatozoa is reached to its final form in the vesiculae only.

The Mucus Glands (Fig. 36)

The mucus glands are paired, large, curved, saclike bodies representing in the drones as accessory glands of general male insects. The mucus gland has an epithelial wall of glandular cells and the muscular sheath has outer longitudinal fibers and inner circular fibers. A third layer of longitudinal fibres is also mentioned by Bishop (1920) in honeybee drones. The secretion of mucus gland is different from that of the spermatic fluid since it is slightly alkaline and coagulates on contact with air and water. Around the mouth of ejaculatory duct there is strongest musculature. The secretion changes its character from fluid to viscus and increase coagulating property to touch or doug by mass.

The Ductus Ejaculatorius

The ejaculatory duct is long, slender and single tube which joins anteriorly to the common point where two lobes of mucus glands meet together at the base. Posteriorly, the ejaculatory duct is connected to the anterior end of inverted penis. There are no opening from the glands into the duct prior to copulation time. However, due to violent contraction of surrounding muscles of the gland the aperture is then formed. This contraction simultaneously brings the mouth of vasadeferentia in contact with the newly formed aperture which permits spermatozoa into the penis of drone through the discharge of mucus from the gland. In the bulb of penis, all spermatozoa are received before mating. This process is completed in the drone before reaching the age, 1 day old; after adult emergence. Thus, with the help of mucus gland (mucus) a sperm pocket is kept ready for discharge before mating.

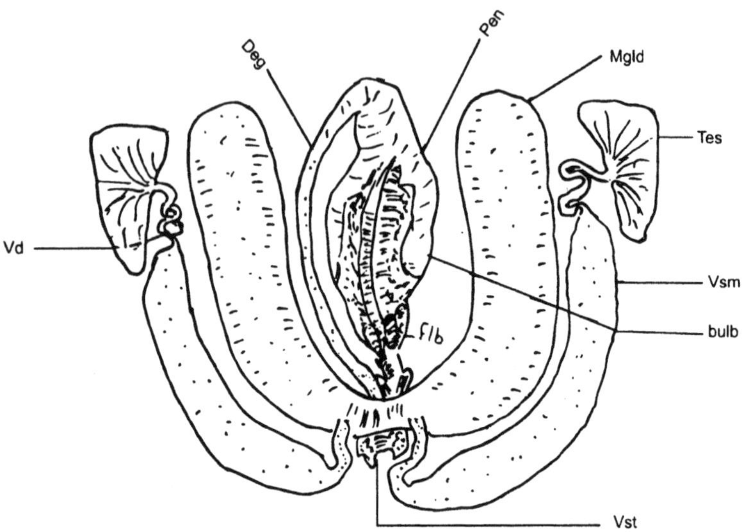

Fig. 36 : Reproductive system of drone bee

Deg	–	Ductus ejaculatorius	Mgld	–	Mucus gland
Bulb	–	Bulb of penis	Vst	–	Vestibulum of inverted penis
Tes	–	Testis	Vsm	–	Vesicula seminalis
Pen	–	Penis	Flb	–	Fimbriated lobe of penis

The Penis (Fig. 37)

The penis of drone is highly developed, endophillic structure in functional form. The ectophillic parts of the genitalia refers to a pair of penis valves and a pair of small, setigerous lateral plates. The penis valves covers the phallotreme and the lateral plates covers over lapping basis of penis valves. The uneverted penis fits in the ventral part of the abdomen and extends upto posterior end of segment III. The penis has following important parts namely, phallotreme, cornua, cervix, fimbriated lobe and bulb (Fig. 37).

Phallotreme is exterior opening with a wide, thin walled chamber called vestibulum. Cornua, the pouchlike, large tapering bodies project from the sides of vestibulum. Anterior to the vestibulum penis is narrowed to a slender neck or cervix. Cervix is irregularly compressed part but, when streched it becomes fairly wide tube. On the dorsal of cervix there is fimbriated lobe and along the ventral side is rounded fold crossed by 'V' shaped sclerites. The cervix anteriorly leads to the large terminal expansion of penis

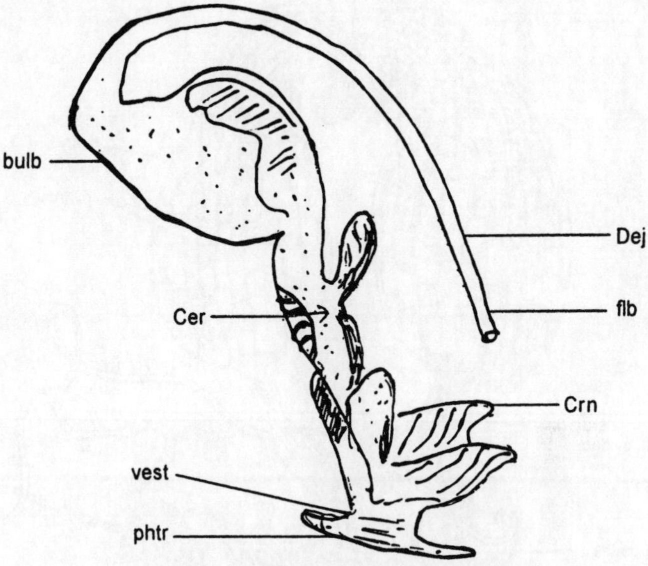

Fig. 37 : Inverted penis of drone bee

called as the bulb.The ejaculatory duct opens into the bulb of penis. The penis bulb has different shapes and size depending upon the presence of content in it. Posteriorly, on the dorsal side there are two large, elongate plates which tapers posteriorly by plates. The structural details are shown in Fig. 38. The inner walls of the bulb are wholly smooth. The bulb lumen opens into the cervix as a triangular aperture.

The Female Reproductive System (Fig. 38)

The female reproductive system consists a pair of ovaries, each with number of ovarioles, a pair of lateral oviducts, a median common oviduct, genital chamber, bursa copulatrix, vagina and accessory glands, etc.

The Ovaries

The ovaries of queen are two huge, pear-shaped masses of egg tubules or ovarioles. The ovarioles are given out from anterior ends of lateral oviducts. The ovarioles are slender and anteriorly tapering threads or tubules. They finally unites anteriorly in a single suspensory strand attached below the ventral wall of the heart. Each ovarioles contains number of bead like eggs and

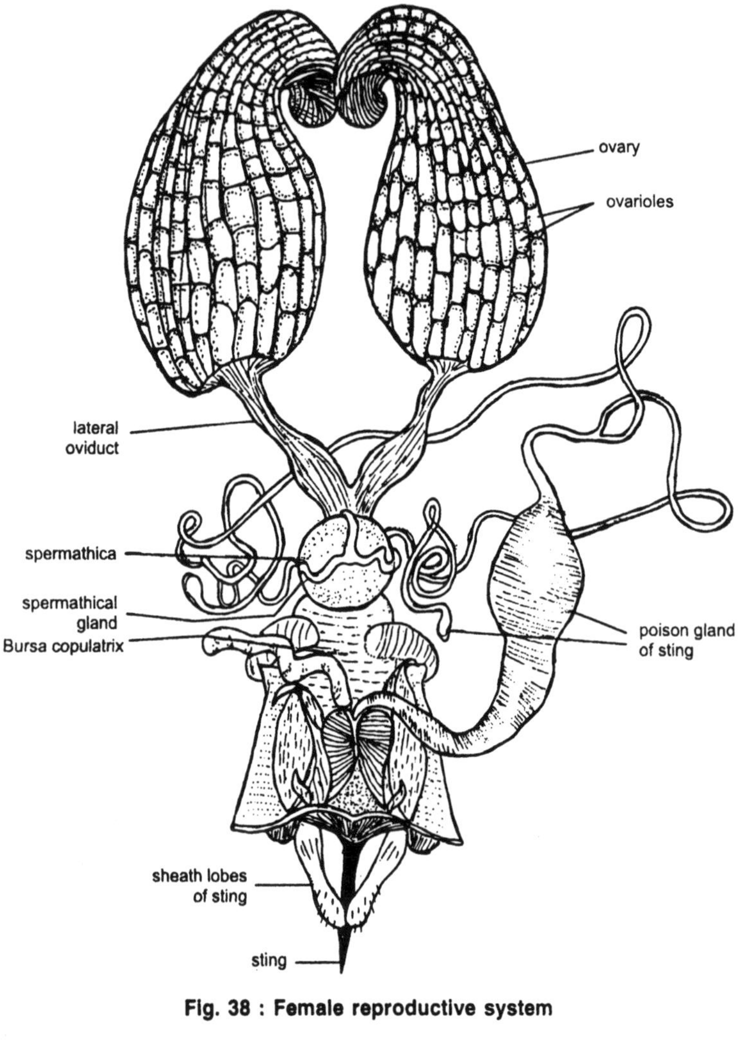

Fig. 38 : Female reproductive system

nutritive cells. Each ovary may contain about 160-180 tubules. The outer wall of each tubule contain thin cellular epithelium. The ovarioles contain multinucleated protoplasmic mass at their extreme upper thread part but, little below the cell boundaries are distinct and the tubules are with solid row of small cells. These are called as primary oogonia, still further the oogonia break up into a mass of polygonal cells which are irregularly distributed.

Here the diameter of tubules get increased. Other small sized cells are also located in the epithelial wall. Oogonia are larger cells in lower level. They are distributed along the axis of ovarioles irregularly. More numerous interspersed small cells are also located. The superficial cells are now found enlarged in forming a definite outer layer just within the epithelium. The large axial cells develop into egg cells called oocytes.

The nurse cells are the intervening small cells, get consumed by growing oocytes as food thus, called food cells. Follicle cells are derived from epithelium of ovariole.

Eggs are cylindrical, curved and rounded at both the ends and measures about 1.53 mm to 1.63 mm in length. The egg has micropile. The matured eggs decends in follicle and new oocytes are generated in upper region of ovariole. The queen lays about 3000 eggs in a single day. During a single year she can lay about 200,000 eggs with an average of 1500 eggs per day.

The Oviduct

There is a single pair of lateral oviducts in female of honeybee. The ovarioles unites posteriorly and join into the lateral oviducts. Further, these lateral oviducts units posteriorly and form a single, common, short wide oviduct. The duct walls are thrown into lengthwise folds which allow expansion. The short common duct is made up of strongly muscular walls and ventral wall shows deep median channel. The common oviduct opens posteriorly into much wider sac like part which is having genital exit called vagina.

The Bursa Copulatrix and Vagina (Fig. 38)

The bursa compulatrix is a wide-open pocket in which biolobed base of sting is freely rested. In this pocket there are three conspicuous openings. Median opening is the entrance to the vagina and the two larger openings goes to a pair of large pouches projecting from the sides of bursa. Each pouch is attached by strong muscle from basal margin of sternum VII. The vigina is an oval sac but, it may vary in form and size with respect to distension. To the ventral wall there is valve like fold projecting into the lumen of spermathical duct. Strong musculature is provided to the fold from sternum VII. Two pairs of slender muscles from spiracle plates of segment VIII are provided to vagina on dorsal wall.

Spermathica

A large, globular sac situated over the vigina is spermathica. The spermathica has its own single paired glands called spermathical glands which opens into spermathica by single tube and spermathica opens into the vagina. Spermathica receives the sperm released by drones into the genital chamber of female and stored for her convenience.

Mating

Multiple mating is common practice of honeybee female (queen). Whether the entire penis including the bulb is everted during mating or only the uneverted bulb is injected into the queen is contravercial aspect. The inverted and reversed position of the penis bulb in the queen suggest that during mating the drone clung to the back of the queen, for this position of the mated pair the bulb, by being merely inserted in the natural position, would be inverted in the sting chamber with its open end directed forward toward the mouth of the vagina (Snodgrass, 1956). According to Zander, in copulation, the penis is everted only as far as the bulb, and only the bulb, which has been torn from the drones body is found in queen. When the penis is fully everted, the ejaculatory duct is drawn clear through the organ and the gonopore is brought to the apex of the bulb. The eversion of penis is produced by the contraction of the abdomen due to having great size of muscles in the drone as compare to worker bee. The supply of spermatozoa received by the queen at one time is generally sufficient to least two or three or four years of the queen's life. However, she can have second mating if needed.

5.
Biology of Honeybee

Honeybee is very useful insect for collection of honey and wax. Biology, ecology and ethology plays a very important role in beekeeping programme. Therefore, knowledge on biology of bee is must. In the present chapter biological peculiarities of honeybee are discussed.

Life Cycle

The honeybee has four stages of its immature forms. These includes egg, larva, pupa and adult.

Egg (Fig. 39a)

The eggs of honeybee (queen) are elongated cylindrical, elongate oval in shape and pearly whitish or pinkish in body colour. The eggs typically tapers toward both the ends and are hymenopteriform. The egg is thicker at one end is the head region of larva. It is slightly curved. The fertilized queen lay her eggs at the bottom of each cell. The eggs found settled on the bottom of egg cell but turns about 45° to its axis after 24 hr, larval formation starts with the third day of egg laying. Egg hatching takes place after 3 days in all castes of honeybee *i.e.*, queen, workers and drones. Queen search empty cells and lay eggs in them. Different kinds of egg cells (caste basis) are prepared by worker bee in hive. The queen can lay more than one egg in a single cell but, worker bee removes all extra eggs and keep only one in each for successful development.

Larva (Fig. 39b, 40 a,b)

The larva is post embryonic developmental event of honeybee. It is typically hymenopteriform which tapers toward both the ends but, posteriorly it is more narrower than anterior. The larva or qrub is leg less and whitish opaque. There are five larval instars in the bee. The young larva of worker bee measures about 1.6 mm in body length. It has semicircular appearence and lies at bottom of the cell. The larva is provided with simple mouth for feeding liquid food but it does'nt show eyes, antennae, legs, wings, sting and reproductive organs. However, appendages are bud like rudiments. Hence, honeybee larvae are not truly apodous and without appendages. Digestive system of larva is relatively simple consisting short and slender stomodaeum, a long cylindrical mesenteron or ventriculus and narrow proctodaeum which opens outside by anus. Tracheal system of larva is also relatively simple.

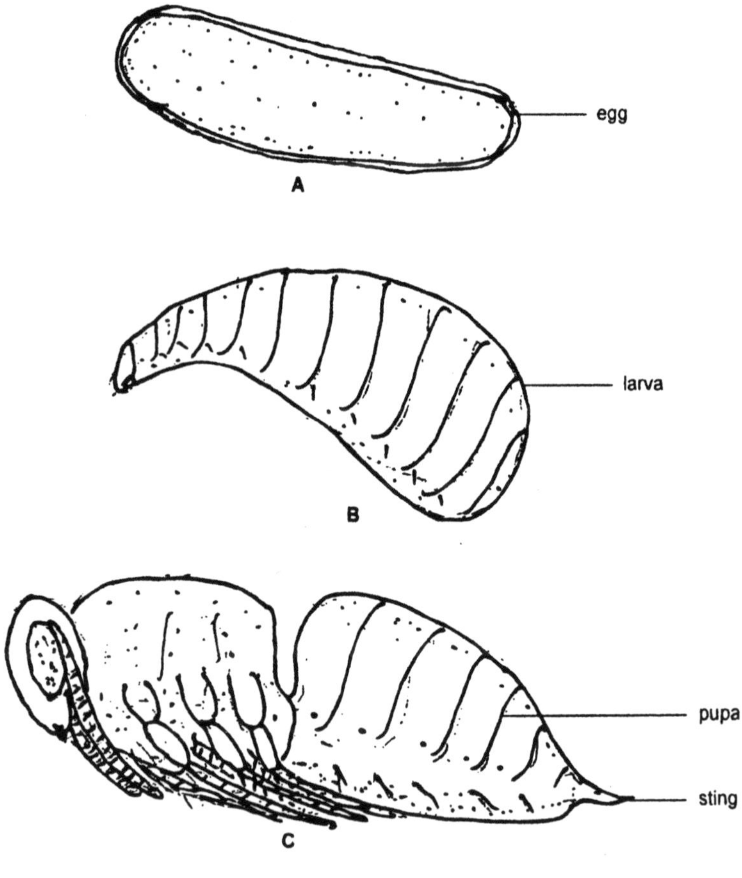

Fig. 39 : (A) Egg; (B) Larva; (C) Pupa

From the main trunk numerous branches of trachae are given out to different body parts. There are ten pairs of spiracles, one on mesothorax, the second on metathorax and other eight pairs on first eight abdominal segments (Fig. 39). The nervous system of larva consists of a brain located in the head, a suboesophageal ganglion which supplies nerves to mandibles, maxillae and labium. Circumoesophageal connectives join the brain and suboesophageal ganglion. Three distinct thoracic ganglia are present with a ventral nerve cord and there are eight abdominal ganglia lie on first abdominal seven segments each. The 8th is larger ganglion located on 8th segment and provides nerves to 9th and 10th segments. On an average the larval durations in worker, drone and queen were 10 days, 13 days and 8 days respectively. The duration for first instar was 24 hr in workers, drones and queens. The same duration was observed in third instar and fourth instars of the bee for above castes. However, 5th instar lasted for 4 days in worker and 3 days in queen while it was 7 days in drones. When the larva full grown, the worker bee sealed the cells with waxy cap.

"Royal Jelly" or "Bee milk" is given to the young larvae of workers, drones and queens. The "Royal Jelly" is stopped for worker and drone larvae after 3rd day by nerse bee but given to queen larvae. "Royal Jelly" is rich in fats and albumens and digested honey and pollen mixed with glandular secretion into the mouth of worker. It is given to queen larva throughout its life time. However, after 3rd day worker and drone larvae are fed with diet of nectary or honey and undigested pollen.

Prepupa

In worker, drone and queen the full grown larva is transformed into pupa within 13 days, 16 days and 11 days respectively. The pupa of honeybee is exarate type. The change for pupa takes place in full grown 5th larval stage, some time before shedding of skin. This stage is called as "prepupa". Some developmental changes takes place in prepupal stage before it moult into pupa. According to Bertholf (1925) the prepupa is freely enclosed in the larval skin. In this case moulting is the separation of the outer layer of the cuticle by dissolving of the inner layer (Snodqrass, 1956). In prepupa head and thorax is more or less adult like but, abdomen is larva like which is not constricted from thorax. Compound eyes are clearly located on the head, mouth parts are not found fully developed. Wings are seen as short wing pads. Thus, the prepupa is with larval skin. This stage lasts for short time, hardly one or two days.

Pupa (Fig. 39c)

Pupal body shows clear distinction of head, thorax and abdomen. The honeybee pupa is exarate type, appendages are not appressed tightly but loosely to the body. Head, antennae, mouth parts, thorax, legs, abdomen and sting, all shows the characteristics of adult form. However, wings are still small as in early pupal stage. Propodeal segment is added to third thoracic segment in pupal stage. Once all the parts develop, pupa does not grow or change in shape during the rest of the pupal period but, changes in colouration of various parts. In newly formed pupa all parts are whitish or opaque coloured but, in matured pupa many parts become brown and black resembling the adult. Adult organs are formed and larval organs withdrawn in pupal stage. Muscles and other body parts are developed which are related to adult life. Thus, before shedding cuticle of pupal form full development of wings and external clothing of hairs takes place. Pupal development takes place in larval cell of comb. No cocoon is found in the case of honeybee. Thus, pupa is located in the larval cell and the cells are sealed by the bee after formation of pupae. Pupa does'nt take food. Fully developed adult bee comes out from larval cell of hive. Newly emerged adult bee just stretch its legs, extends antennae, wings are spread and becomes ready for doing the duties assigned to, if the caste is worker but, if it is queen she is preceded and liquidated in its cell by the first queen. However, drone is nothing to do in the hive except feeding and fertilizing female when she call for his services. After mating, oviposition is started by the queen after 3 days only.

The pupal periods found in worker, drone and queen are 8 days, 8 days and 4 days respectively. Thus, total life cycles, from egg to adult in different castes of honeybee requires 21 days, 24 days and 15 days respectively. The honeybee is complete metamorphic insect *i.e.,* wings develop internally, pupal stage present, larva and adult is disimilar, etc.

During the development of honeybee, nurse bee takes care of larvae with respect to type and quality of food, feeding time, and proper growth of the larvae and pupae of different castes. During the larval development nurse bee visit a cell for about 1300 times. Thus, proper growth and development of bee is possible if appropriate quality and quantity of food is provided by nurse bee. On an average worker, drone and queen consumes 1.5 mg, 1.045 mg and 3.25 mg of royal jelly respectively during their larval stages.

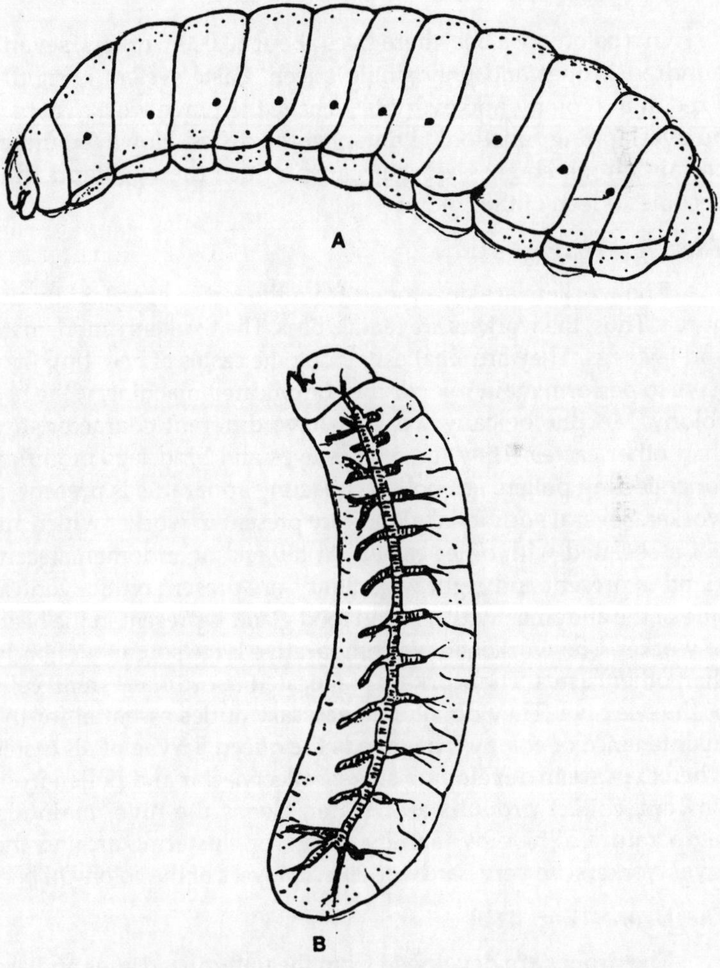

Fig. 40 : (A) Matured larva with spiracles; (B) Tracheal system in matured larva

Caste System in Honeybee

Honey bees are social insects, therefore there is division of labour, caste system and polymorphism. The castes of bee consists of:

1. Workers,
2. Drones, and
3. Queen.

In a colony of a bee, there lives about 60,000 workers, several hundred drones and only single queen. Caste system is rigidly fixed in the colony, and every caste have their own work. There is no overlapping functioning amongst the castes. However, under certain circumstances worker can produce the eggs and thus becomes queen of the colony.

The Worker (Fig. 41a)

The workers are developed from the fertile eggs laid by the queen. Thus, the workers are female bees. The workers cannot mate and lay eggs. They are smallest among the castes of bees but, they have to perform main role for the complicated machinery, the bee colony. Morphologically workers have different characteristics than other castes. They have large eyes and hind legs modified for collecting pollen, *i.e.*, pollen collecting apparatus is present in worker. Several additional glands are present in worker which are not associated with other castes. At the end of abdomen a scent gland is present and eight wax glands are present on the ventral side of the abdomen while, brood food gland is present in the head of worker. The workers have comparative large soft area of body than other castes. The workers are located in ordinary small cells of the bee hive. They do all the necessary duties essential for the maintenance of colony. They feeds the queen, larvae of all castes when they are in developing stage, collect nectar and pollen from flowers, collect propolis and water, cleans the hive, maintain temperature of hive by fanning wings, or clustering around the hive. Workers are very hardworker employees of the colony of bee.

The Drone (Fig. 41b)

The drones are developed from the unfertilized eggs and are male bees. The drone is refered as "King" of the colony. The drones are bigger than workers and smaller than queen in body length.

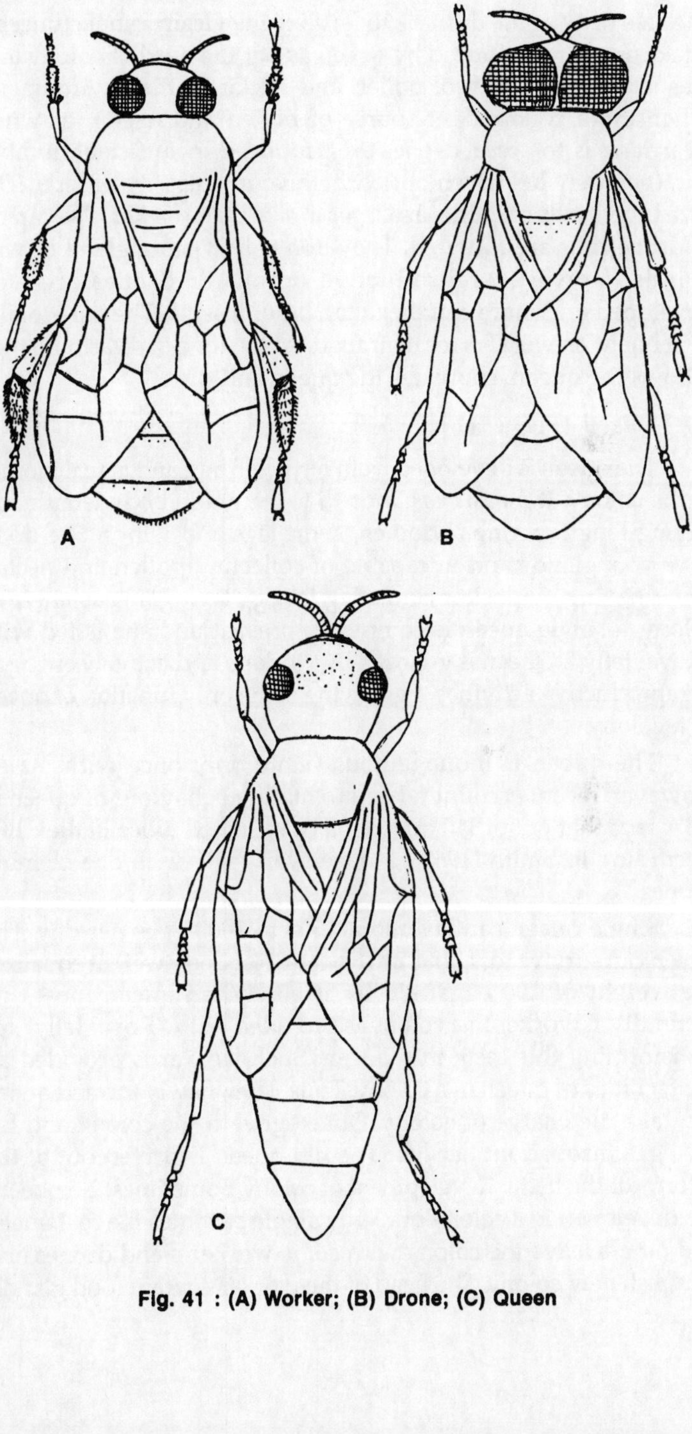

A B C

Fig. 41 : (A) Worker; (B) Drone; (C) Queen

The sole duty of the drone is to fertilize the virgin female (queen). The drones are produced by queen as per the need of colony and season or availability of pollen and nectar in the environment. When there is abundant source of pollen and nectar, or when swarming is to occur, drones are produced in sufficient number but, they may be driven out of the hive in scarecity of food. The hive brood cells of drone are larger in size than worker. The caping of drone cells are rounded. They don't show sting gland or wax glands. However, reproductive organs of drones are well developed and more complex than other insects. They are totally dependent on workers for their food. At the time of swarming they follows the queen, copulates the queen and dies.

The Queen (Fig. 41c)

The queen is developed from fertilized egg and is real mother of the colony. It measures about 15 to 20 mm in body length and having long tapering abdomen, short legs and wings. She don't have wax glands and uncapable of collecting pollen and nectar. She is having ovipositor cum sting for egg laying. In each hive or colony, a single queen is commonly present and she is fed with "Royal Jelly". Queen is guarded by workers and actively engaged in reproductive activities. Egg laying is the only function of queen in a colony.

The queen is monogamous (mate only once with male). However, recent account related to mating behaviour of queen in USA says that out of 110 queens 55 have mated twice in their life. According to Smith (1963), a queen can mate with one or more drones.

Single queen can lay about 1500 to 2000 eggs per day. The queen can survive for about 2 to 5 years. Her body weight equalizes the weight of 100 eggs laid by itself. When queen looses its fecundity, a worker in a colony is provided with "Royal Jelly" for transforming this caste into queen. Such worker is provided all the facilities of queen in a colony. Thus, later, newly formed queen can take the charge of colony. Queen govern the colony and lay the eggs throughout her life. The old queen is driven out of the colony for the better development of colony. Sometimes, 2-3 queens are developed in a colony but, only single becomes the real queen and others leave the colony with some worker's and drones and establish new colony. The head of the worker contain food glands.

The workers touch the queen with their antennae, the two sensitive feelers lick her. In this way, they get queen substance from queen which is then passed to other bees in colony.

Development of Queen

Workers build a special queen cell on the face of comb for the development of queen in that particular cell. Such cells are called as supersedure cells. This is done if the queen becomes old or sick or less fecund. The queen lays an egg in each cell. These eggs hatched into white, leg less qrub within 3 days. The newly emerged larva is fed with brood food secreted by the head food gland. The larva floats in the liquid food thus, excess food is provided to larva within 5 days, the larva is full grown. After that the worker bee seal the cell by thin cap of wax. The full grown larva then spin cocoon and pupate inside. During pupal stage the creature slowly changed into a queen. Pupal stage lasts for 7-8 days. Thus, within 15-16 days queen is developed from egg laid by old queen. The newly formed queen is virgin queen which emerge from the queen cell by biting and opening cap of the cell. Newly emerged queen feed on honey and flies out of the hive for few days for mating purpose. She mates with one or more drones and rejoins the hive and within 2 days after mating, she start laying the eggs in the hive and thus, it take the charge of hive.

Both old and new queens can live together for a week or two. During this period the old queen dies and the dead queen is thrownout by the worker bees of the colony. In case of sudden death of queen or queen taken away, the workers rear female eggs larvae (young, 1-2 day old) providing same food to the larva of worker and queen for first 2 days. The developing eggs/larvae of proposed queen is in worker cell but, the bees enlarge the cell into queen cell. The cell thus formed is called as emergence queen cell. The proposed queen larva is fed with "Royal Jelly" as like larva developing in supersedure cell which then will be develop into queen caste.

Queen has two kinds of eggs, female eggs (fertilized) and male eggs (non fertilized). The queen receives the sperms from male (drone) and store them into the spermathica for life long. The queen fill the eggs in worker's cell or drone cells as per her need. If she want to have drones she select non-fertilized eggs and laid in drone cells while, female eggs (fertilized) are laid in worker cells for producing workers. Queen is also produced by fertilized egg. Laying of workers resulted in formation of drone as eggs are not fertilized in this case. Sometimes, worker may lay female eggs but are unfertilized. The capping of laying worker cells are rounded like the drone cell caps.

Hive or comb is the house of honeybee which consists of hexagonal cells made up of wax secreted by wax gland of worker. The hives of bees are found hanging vertically from branches of trees, building, rocks, crevices, etc. Thousands of hexagonal cells are arranged in two opposite rows on a common base in each hive or comb. Propolin *i.e.,* a resins and gums secreted by plants is used for reparing of the hives. The comb consists of upper part of cells containing honey and lower part containing brood cells. The broods are of three kinds namely, *Worker cells* (Fig. 42) for workers, *Drone cells* (Fig. 42) for males (drones) and *Queen cell* (Fig. 42) for queen. The drone and worker cells can be reused by bee but, queen cell is not reused by the new queen. However, there are no any special cells for lodging adults. Therefore, adults generally found

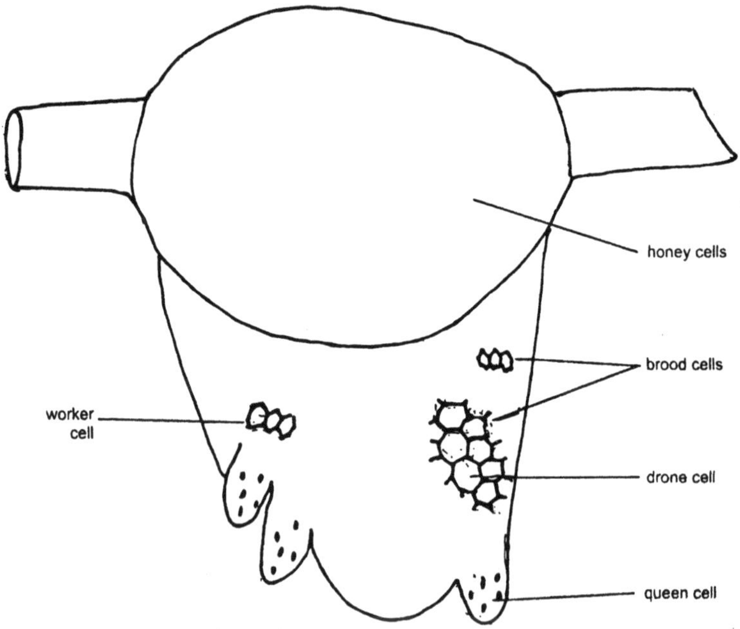

Fig. 42 : Honey comb with different cells

clustering/moving around the comb. The combs are built parallel to each other with their centres placed about 1¼ in part. In hot region of India they are little closer. Worker cells are smallest, 1/5 in across and drone cells are larger, about 1/4 in across.

The Food of Bee

The honeybee feed on pollen and nectar. Nectar is sugary liquid derived from flowers of varieties of plants including trees, shrubs and herbs. The nectar collected by honeybee with the help of tongue is kept in honey stomach and carried to comb. It is then put into the comb cells and converted into honey. During the process excess water is evaporated and enzymes are added to it so that most of the sucrose in nectar is converted into dextrose and levulose. Pollen are also collected from several flowers. The pollen grains attached to body hairs of bees are collected by pollen baskets of hind legs. Once the baskets are filled with pollen grains, the bee returns to comb or hive and vacate the baskets by poreing pollen grains into cells. Other bees backages the pollen grain properly later on. The bee again go for field visit for collecting pollen and nectar.

The bees have ability to collect their food from longer distance but, mostly obtain their food from 3/4 mile of the comb. The bees live on honey diet alone for several months. For production of brood food and production of continuous wax the bees should eat pollen.

Food and Dance Language

An Australian scientist Dr. Karl Von Frisch, Professor of Zoology, University of Munich, Germany investigated language of bees in the early 20s. After several times of observation, Frisch reported that forager bee on return to the nest made two types of dances on the surface of bee hive namely, "Round dance" and "Tail-wagging" or figure of eight dance. The round dance is used to indicate a short distance, less than 50 m by the bee, *A. mellifera*. The bee moves in circles, first in one and then in opposite direction, clock and anticlockwise (Fig. 43a). The tail-wagging dance was used by the bee to indicate a longer distance of food, beyond 50 m. During this dance, the bee made two half-circles in opposite directions with a straight run inbetween (Fig. 43b). The bee shaked (waged) its abdomen from side to side in straight run and the number of wags per unit time was related to the distance of food located. More number of wags was related to shorter distance and less number to farther the food. The phenomenon was also studied in the species *A. indica*. In *A. indica* 10.5 wags in 15 seconds were related to the distance of 50 ft and 4.4 wags to 1000 ft. Thus, different species have different account. According to some workers the sound produced during tail waged dance have also the relation with the language of bee.

For the foraging bee reaching to target of food needs clue for direction. This is conveyed by the angle (Fig. 43c). The angle between the straight run and the sun shows the direction of the

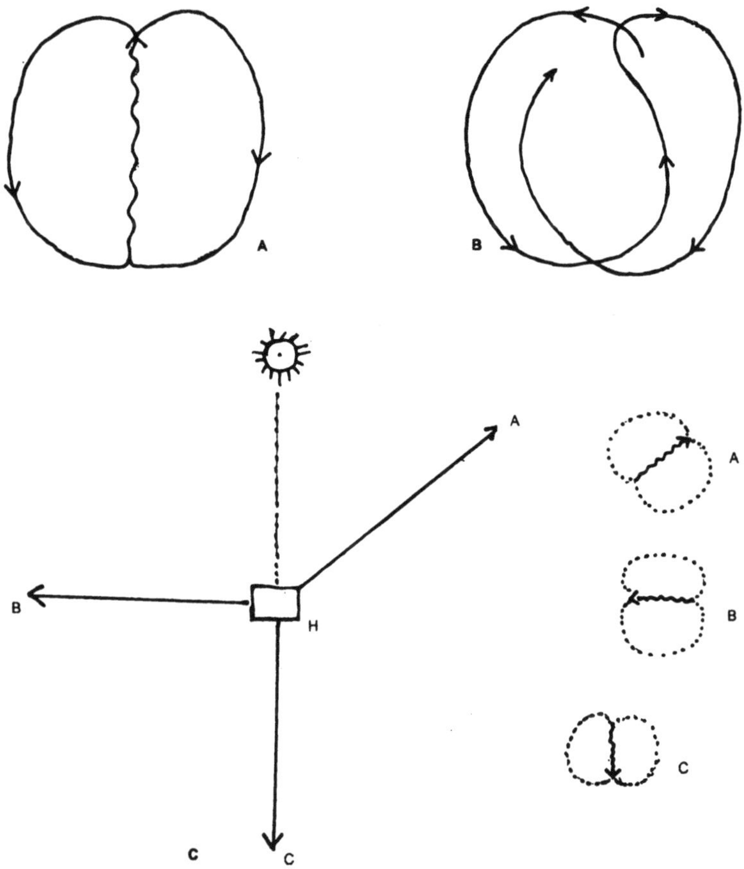

Fig. 43 : (A) Wag-tail dance; (B) Round dance; (C) Sense of direction for outgoing forager

source of food. Bees have correct judgement of sun even at clowdy environment. The bee dances straight up the comb if the source of food lies in the same direction as the sun. The bee faces straight down the comb if the source of food lies on the opposite side of the comb from the sun. The bee dances with the top of the hive on its right if it has to fly with the sun on its right side.

Maintenance of Proper Conditions in Hive

The bee can have small amount of heat in their body. In colder condition, the bees cluster together and increase the heat in comb. If there is compact clustering on the comb by bees, heat will be produced more and if clustering is loose, less heat is generated in the comb. Thus, in cold season the·bees warmup the comb by clustering together and in hot weather the comb temperature is being decreased by clustering loosely on the combs. During rearing of broods the bees maintain the temperature of hive from 88°F to 95°F. In conditions when cold wind is blowing or water droping on the hives, the bees on the outside cluster becomes more compact and stored honey and also keep the bees warm. Therefore, bees are not affected by sudden fall in temperature, at night time temperature of comb may falls suddenly.

Honey, water and nectar plays important role in decreasing the temperature of comb. The bees maintain the temperature below 97°F by fanning the hive and circulating. Honey take long time for warmup, thus honey keep the hive cool and help in decreasing temperature of the hive. In hot weather water is brought to the hives by bees which is evaporated and the phenomenon is helpful for cooling the hive. In nectar sufficient quantity of water is present which is again helpful for cooling the hive by evaporating water present in them. When temperature increases beyond the range the bees becomes bad tempered. Therefore, the hives are kept in shaded place and saved from the sun. Hive box also painted white in certain hot countries. This is helpful for keeping better temper of the bees and further for producing more honey.

Ventilation to bee hive, proper food to bees, and availability of flora have great importance in bee keeping business. Particularly in cold weather countries, ventilation to hive is must. In the hive, bee needs fresh air for breathing. Generally, the bees want to get rid from stale air of hive. Stale air contains much more quantity of water which rises at top of the hive. There should be an outlet at top of hive for ventilation of air. The moisture at hive top condenses on the cover, outer combs, and the walls of the hive resulting in unhealthy conditions to bees, form moulds. Proper ventilation and sufficient food encourages to the longer life of bees in coldest season. However, hive must not be exposed to the winds which chills the bees. Secondly, the bees die more quickly than should in such conditions and also consume more food in such chilled conditions. The bees should be sheltered from winds from all directions. There should be proper size and place for entrance to hive so that stale air can passout and fresh air can come in the hive but, prevent enemies of bees to rush in the hive.

Antenna is the sense organ for bees. With the help of antenna, bees can test the smell. Antenna is sensitive as mouth and nose of humans in bees. Antenna is also organ of sense of touch. Richer nectar are discriminated by bees in the field on the basis of strong and weak sugar solutions. The bees find out flowers quite easily because they identify small differences in scents and mixture of scents of flowers. The bees have their own ability to identify the difference between their own hive and hive of other bees. Likely, they also detect water from distance and the natural enemies too.

The eyes of bees are very sensitive to movements therefore bees get protected from their enemies. Bees can see the flowers waving in the nature. They also see ultraviolet colour but can not distinguish black and red colours. Likely, they can not distinguish between orange, yellow and green, but very interestingly distinguish any of these colours and blue or blue-green. Therefore, bees can see flowers in different colours than the man. According to Smith (1969) a purple flower does not look purple to a bee. Purple is made up of red and blue and bees can not see red. Therefore, the flower will look blue to the bee and might be tinged with ultraviolet as man can not see the ultraviolet. Painting the fronts of hives with colours, black (or red), white, yellow or blue will be helpful to bees for distinguishing their own hives.

Swarming

It is believed that swarming in the bee is occured for reproducing the colony. Swarming is defined as the process of leaving off the colony by the queen. Swarming occured at the end of spring or early summer. However, the real cause of swarming is not cleared upto date. When a colony becomes very strong by population (broods) and when there is plenty food material in the environment, the bees build few queen cells generally at the bottom or edges of the comb. Queen lay her eggs in such cells and new queen is developed. During the development of such broods special food, "Royal Jelly" is given to individual. When old female becomes less fecund, fewer and fewer eggs are laid and finally the old queen cell is sealed and the old queen can be driven or she can fly from the hive at any time with good number of old workers and further settle down on branch of a tree or other suitable place. The queen generally leave the hive at fine fore noon. The new empress of the hive gives the order to kill other sisters in the colony when it take the charge of colony. In another concept, when old queen leaves the comb, a worker is given "Royal Jelly" and the worker is then converted into fertile queen. Newly formed such type of queen again never tolerate the other sisters in the colony and she order to kill them. Newly emerged queen may go to other queen cells and destroy them and kill the sisters. Some times newly formed queens live together for time being and separate out for formation of new colony.

Scout bees are involved in swarming initiation. They decide and take trials for swarm where to go. They find out place of settlement and come back and take other castes to the target place of hive to be constricted. They guide by making typical dance form second swarm called cast-several castes may come out until only one virgin queen is left.

Abscounding

Abscounding is migration of the complete colony from one place to another due to certain unfavourable conditions like scarcity of nectar (required flora), destruction by natural enemies such as termites, ants, wax moths, etc. However, abscounding is different from the swarming phenomenon. It is complete desertion of a hive. Before abscounding bees drink all honey present in comb. Thus, they leave empty comb and migrate to other place. Abscounding may occur even due to the excessive disturbance from beekeeper. However, abscounding can easily be prevented by fullfilling the (urgent) needs of bees *i.e.*, water, surgar solution as food, ventilation to hive, protection from natural enemies, etc. During abscounding, the swarm may be joined to the another swarm from another colony. Such swarms may mix with each other peacefully or may have battle among them. However, they can not establish a colony till the queens get reduced to single in a colony. The fighting behaviour is not seen in reproductive swarms. The bees can be rehived by providing food immediately. *A.mellifera* in cold climate do not abscound in adverse condition (Scarecity of food) but die slowly in want of food. Abscounding frequency within the abscounded colony have also studied by some workers. It is observed that reabscounding in certain species is possible within a few hours or a few days.

Supersedure

If a queen suddenly dies or when egg laying capacity of the old queen is lost, a new young queen takes the charge of the old queen is called supersedure. When a queen lays a high proportion of male eggs, supersedure queen replaces the individual queen. Queen substance is diminished in such queens that stimulate the worker bees to build supersedure cells, which are 2-3 in number and larger in size and made in the middle of the nest. The excess queens are killed and the young queen starts egg laying after mating with drones.

Nuptial or Marriage Flight

Swarming is of two types. In first case it is with old queen and in second case it is with 7 day old virgin queen. The swarming led by young queen is followed by drones. This is refered as marriage flight. The queen mates and copulates with one of the drones in the sky and get fertilized. The drones dies after copulation. The spermatophores are received by the queen and stored in spermathica. The copulated pair falls on the ground, drones found died but the queen find her way to hive for taking the charge of hive weather and flowering plants have the relationship with the activity of the honeybees. When bee can not collect pollen or nectar, the period is called as the *"period of dearth"* or *"dearth season"* and when pollen and nectar are available the period is called as *"buildup perioa"* or *"buildup season"*. In temperate climates, the period of dearth is winter and in tropics during dry season. The duration of heavy rainfall is also the dearth period of bees. During this period queen lays less eggs and fecundity is reduced to its minimum level. Sometime egg laying completely stopped. In such period bee spend most of their time in the comb leaving empty combs.

In temperate climates, build up period occurs in spring, in the tropics it occurs just before the rains, during light rains, or after the heavy rains. During this period the bees collect large amount of pollen and nectar as much as possible. The bees also collect much more quantity of water. Egg laying is started by queen and more and more bees are produced the colony is expanded to its peak and colony flow becomes ready to fly. *"Honey flow period"* is the period during which the bees collect more nectar than they needed for day to day life. The bees store the surplus nectar in the form of honey. During this period the brood hive constricts and honey flow is used for developing the broods and unable to store any surplus honey. The honey flow is possible in temperate climates. While in tropics at the beginning of rains or after the end of rains. Harvest period is visualized after the end of honey flow in autumn or fall in temperate climates. Brood rearing is decreased in this period since little pollen and nectar is collected due to unavailability. However, any surplus is stored in the combs of the brood nest, cells vacated by emerged broods. During a year the bee has following periods:

1. A period of dearth,
2. A buildup period of 6 to 12 weeks,
3. A honey flow period, and
4. A harvest period.

During a single year more than one honey flow periods are possible with or without gap in them.

Flora of Apiculture

Apicultural flora is of two types:

1. Wild flora, and
2. Cultivated flora.

A very large number of wild flora is visualized as good source of pollen and nectar for apiculture purpose. Jamun, acasia, ber, berseem, boxelder, apple, brassica, sunflower, safflower, mustard, turnip, cabbage, carrot, coconut palm, almond, apricot, etc. are important flora for beekeeping business.

References

Bailey, L. 1954. The respiratory currents in the tracheal system of the adult honeybee. *Journ. Exp. Biol.* **31,** 589-593.

Bertholf, L.M. 1925. The moults of the honeybee. *Journ. Econ. Ent.,* ***18,*** 380-384.

Betts, A.D. 1923. Practical bee anatomy. 85 pp., The Apisclus, Benson, England.

Bishop, G.H. 1920. Fertilization in honeybee. The male sexual organs. Their histological structure and physiological functioning. *J. Expt. Zool.,* **31,** 225-265.

Butler, C.G. 1955. The role of "queen substance" in the social organization of a honey bee community. *American Bee Joun.,* ***95,*** 275-279.

Casteel, D.B. 1912. The behaviour of the honeybee in pollen collecting. US Dept. Agri., *Bur. Ent. Bull.,* **121,** 36pp.

Day, M.F. and Waterhouse, D.F. 1953. Functions of the alimentary system. In Roider, Insect Physiology. Chap. ***11,*** 288-310.

Duncan, C.D. 1939. A contribution to the biology of North American Vespine Wasps. Stanford Univ. Publ. *Biol. Sci.* ***8,*** 272pp.

Flanders, S.E. 1950. Control of sex in the honeybee. *Scientif Monthly,* ***71,*** 237-240.

Fyg, W. 1952. The process of natural mating in the honeybee. *Bee World,* **33,** 129-139.

Haydark, M.H. 1943. Larval 'food' and development of castes in the honeybee. *J. Econ. Ent.,* ***36,*** 778-792.

Kenyon, F.C. 1896. The brain of the bee. *J. Comp. Neurol.,* ***6,*** 133-210.

Kerr, W.E. 1951. Sex chromosome in honeybee. *Evolution*, **5**, 80-81.

Mackensen, O. 1943. The occurrence of parthenogenetic females in some strains of honeybees. *J. Econ. Ent.*, **36**, 465-467.

Manley, R.O.B. 1948. Workers mating. *American bee Journ.*, **88**, 305.

Manning, F.J. 1949. Sex determination in the honeybee. The Microscope, May-June, 1949, 5pp.

McIndoo, N.E. 1914. The olfactory sense of the honeybee. *J. Expt. Zool.*, **16**, 265-346.

McIndoo, N.E. 1922. The auditory sense of the honeybee. *J. Comp. Neurol.*, **34**, 173-199.

McIndoo, N.E. 1945. Innervation of insect hearts. *J. Comp. Neurol.*, *83*, 141-155.

Morison, G.D. 1928. The muscles of the adult honeybee (*Apis mellifera* L.) Part III. *Quart. J. Micr. Sci.*, **71**, 511-526.

Nelson, J.A. 1915. The embryology of the honeybee. 282pp. Princenton.

Nelson, J.A. 1924. Morphology of the honeybee larva. *J. Agri. Res.* *28* (12), 1167-1213.

Oertel, E. 1930. Metamorphosis in the honeybee. *J. Morphol.*, **50**, 295-340.

Park, O.W. 1954. How bees make honey. *American Bee Journ.*, **94**, 296-298.

Roberts, W.C. 1944. Multiple mating of queen bees proved by progeny and flight tests. *Glean. Bee. Cult.*, **72**, 255-259.

Rodionov, V.V. and I.A. Shabarshov. 1983. The facinating world of bees. Mir Publ. Moscow.

Schmieder, R.G. 1928. Observations on the fat body in Hymenoptera. *J. Morphol.*, **45**, 121-184.

Smith, F.G. 1963. Beekeeping. Oxford Uni. Press, Toronto.

Snodarass, R.E. 1956. Anatomy of the honeybee. Comstock Publ. Asso. London.

Shrivastava, K.P. 1993. A textbook of Applied Entomology II. Kalyani Publ., New Delhi.

Weaver, N. 1955. Rearing of honeybee larvae on 'Royal Jelly' in the laboratory. *Science*, **121**, 509-510.

Willson, R.B. 1955. Royal Jelly : A review. *American Bee Journ., 95,* 15-21.

Woyke, J. 1955. Multiple mating of honeybee queen (*Apis mellifera* L.) in one nuptial flight. *Bull. Acad. Poloniaise Sci.* Cl. II, *3*(5), 175-180.

INDEX